The New York Times

Book of

INSECTS

The New York Times

Book of

INSECTS

EDITED BY

NICHOLAS WADE

THE LYONS PRESS
Guilford, Connecticut
An imprint of The Globe Pequot Press

The Lyons Press is an imprint of The Globe Pequot Press

10 9 8 7 6 5 4 3 2 1

Printed in the United States of America

Designed by Joel Friedlander, Marin Bookworks

Library of Congress Cataloging-in-Publication Data

Science times book of insects.
    The New York Times book of insects / edited by Nicholas Wade.
        p. cm.
    Originally published in 1998 under the title: The Science times book of insects.
    ISBN 1-58574-792-0 (pbk. : alk. paper)
        1. Insects. I. Wade, Nicholas. II. Title.

QL463.S36 2003
595.7—dc21

2003047409

Contents

2   Attack and Defense

3   Rituals of Insect Courtship

Introduction

TINY, PUNY, PESTILENTIAL. Apart from a few exceptions like bees and butterflies, insects have a definite image problem. Zoologists may respect them as the most successful class of arthropods, but to most people they are just bugs, menacing nuisances to be swatted or stamped on.

For anyone who seeks to understand nature, this view of insects is seriously flawed. A handful of insect species have become pests, but the overwhelming proportion are beneficial. This is not because insects bear any positive sentiment toward humankind, merely that the planet's terrestrial life systems could not operate without them.

Insects have an evolutionary pedigree that stretches back some 300 million years. There is one living species of human, but insects have radiated into more species than anyone can count. Just over a million species have been described and named by scientists. Estimates made on the basis of samples put the total number of insects species at between 15 million and 30 million.

Over the course of evolution, insects have learned to live in every extremity of the earth, from the scorching sands of the Sahara to the snow-mantled wastelands of Antarctica. In terms of biomass, even the human population is no match for invertebrates: The weight of all Americans is less than one fiftieth that of the insects, earthworms and spiders that live in the United States.

Because of their long pedigree and vast numbers, insects have come to play key roles in every terrestrial ecosystem. Without insects, there would be no flowering plants or trees. Without insects, trees would not rot for decades, animal wastes would not get speedily recycled.

Not surprisingly, the little creatures that play such a grand role in global ecological affairs are miracles of compact and efficient design. The nature of their exterior skeleton has kept them small, but within this constraint many

1

species have contrived extraordinarily inventive ways of life. There are leaf-cutting ants that have mastered the art of agriculture and raise crops of a mushroomy fungus in underground chambers. There is the dung beetle, whose cleanup efforts save the planet from becoming a pigsty. There is the cockroach, an insect whose only fault is that it has cunningly adapted itself to an urban lifestyle.

Since most insects perform their labors out of sight, few people appreciate the accomplishments of these miniature animals in mastering the complex tasks necessary for their survival. It is hard to improve on a compliment from an unexpected source, the great historian of the Roman Empire. Edward Gibbon, after describing the wonders of St. Sophia, the soaring basilica constructed in Constantinople by the Emperor Justinian I in A.D. 537, offered this reflection: "A magnificent temple is a laudable monument of national taste and religion, and the enthusiast who entered the dome of St. Sophia might be tempted to suppose that it was the residence, or even the workmanship, of the Deity. Yet how dull is the artifice, how insignificant the labour, if it be compared with the formation of the vilest insect that crawls upon the surface of the temple!"

Huntsman spider feeds on household flies and roaches.

The Importance of Being Spineless
Invertebrates, animals that lack backbones, are so pervasive that it may be said life depends on them. The most numerous are insects, but there are also spiders, lobsters, corals, worms, etc.

Though Gibbon was as interested in diminishing St. Sophia as in praising the insect world, his tribute was not misplaced. The more biologists understand about insects, the more they find to admire in their skills, their behaviors and their survival strategies.

It's not hard to suppress one's admiration for mosquitoes and houseflies. But most of the other insects that cross one's path are innocent wanderers that want only to return to their unpresuming microworld. The next time you find a katydid or a damsel fly has strayed into your house, don't swat or stamp on it. Put a glass over it, slide an envelope or stiff piece of paper underneath, and release the little creature back into the wild where it belongs.

The articles selected here, originally written for the science section of *The New York Times,* report biologists' new findings and understandings

Some insects prey on invertebrates humans dislike.

Searcher beetle preys on harmful caterpillars.

Many insects feed on other invertebrates, sometimes specializing in one species, and many of the prey species are crop pests.

"The little guys who run the world"
Invertebrates build soil into the fertile matrix for plant life, and therefore support life for all mammals, including humans. Some insects act as the matchmakers, for much of plant sex life, as many plants rely on insects for pollination. They also do important work in seed dispersal.

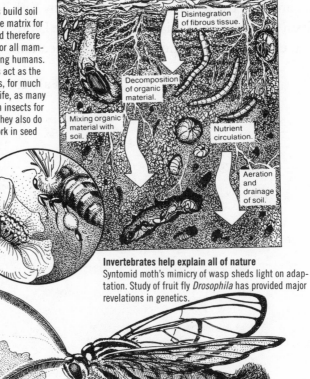

Disintegration of fibrous tissue.

Decomposition of organic material.

Mixing organic material with soil.

Nutrient circulation.

Aeration and drainage of soil.

Invertebrates help explain all of nature
Syntomid moth's mimicry of wasp sheds light on adaptation. Study of fruit fly *Drosophila* has provided major revelations in genetics.

3

Dimitry Schidlovsky

about insects. In being collected together they gain a coherence that is lost in newspaper presentation.

Zoologically minded readers will note that several spiders, a scorpion or two and some mites have slipped in among the insects in this book. These creatures belong to the class known as arachnids and of course have eight legs, unlike their cousins the insects, which have six. It seemed reasonable to include them in a book on insects because they have a similar form and lifestyle, and are often lumped in with insects in people's minds. This conflation is not without basis, since insects and arachnids are both members of the arthropod phylum, the grand group of animals that have jointed legs and hard outer bodies in place of an interior skeleton.

The usual life of a newspaper article is even less than an insect's. My colleagues and I on the science section of *The New York Times* are grateful to Lilly Golden for the idea of this book and to the Lyons Press for giving our writings an extended life in the form of this collection.

—NICHOLAS WADE, Fall 1998

1

MAESTROS OF EARTH AND AIR

nsects are an ancient order and in the course of evolution many species have developed extraordinary capabilities.

The monarch butterfly, for example, migrates to North America east of the Rockies from its overwintering sites in the mountains of Mexico. The butterflies that return to Mexico in the fall are not the same insects, but their descendants, several generations removed. It is not known how knowledge of the ancestral home is transmitted across the generations.

The butterfly is adept at chemical warfare as well as navigation. Its caterpillar stores a poison made by its foodplant, the milkweed, and the butterfly is so distasteful to birds that few will attack a monarch twice.

The leaf-cutting ant is another virtuoso of the insect world. It cuts up leaves and chews them to a fine mulch on which it grows a special fungus related to mushrooms. The ants' fungus gardens are underground, in a collection of chambers whose combined volume can equal that of a living room. The ants live in colonies, up to 8 million strong, with castes of several different sizes and occupations.

Insects may be small and weak, but these tiny creatures have remarkable capabilities. The following articles profile some of the maestros of the insect world.

In Recycling Waste,
the Noble Scarab Is Peerless

IN THE VAST WORLD OF BEETLES, they have the stamp of nobility, their heads a diadem of horny spikes, their bodies sheathed in glittering mail of bronze or emerald or cobalt blue. The ancient Egyptians so worshipped the creatures that when a pharaoh died, his heart was carved out and replaced with a stone rendering of the sacred beetle.

But perhaps the most majestic thing about the group of insects known romantically as scarabs and more descriptively as dung beetles is what they are willing and even delighted to do for a living.

Dung beetles venture where many beasts refuse to tread, descending on the waste matter of their fellow animals and swiftly burying it underground, where it then serves as a rich and leisurely meal for themselves or their offspring.

Each day, dung beetles living in the cattle ranches of Texas, the savannas of Africa, the deserts of India, the meadows of the Himalayas, the dense undergrowth of the Amazon—anyplace where dirt and dung come together—assiduously clear away billions of tons of droppings, the great bulk of which comes from messy mammals like cows, horses, elephants, monkeys and humans.

Scientists have long appreciated dung beetles as nature's indispensable recyclers, without which the planet would be beyond the help of even the most generous Superfund cleanup project. But only recently have they begun to understand the intricacies of the dung beetle community and the ferocious interbeetle competition that erupts each time a mammal deposits its droppings on the ground.

The Industrious Cycle of Beetles

Millions of energetic dung beetles function as a global sanitation crew. As many as 16,000 can be found in a single elephant patty. Shown here is *Scarabaeus sacer,* the Egyptians' sacred scarab, found in North Africa and Egypt; it consumes the dung of large herbivores, like cattle, and uses it to protect and feed its larvae.

Rolling a Brood Ball

A beetle forms dung into a ball that can be rolled to the nest, some distance away in an earthen tunnel. Eggs are laid in the ball, where the larvae hatch and incubate for months.

Improving the Soil

The beetle's digging activity aerates the soil, allowing oxygen and nitrogen to penetrate deep down and increasing the growth of plants.

The dung microhabitat

The dung of herbivores is extremely nutritious for beetles, so many compete for it. Beetles are so active that they will carry off a cow patty in less than an hour. The success of beetles in monopolizing dung means fewer pest insects, like flies. The beetles have many strategies for getting their share of the dung heap, including burrowing, rolling and stealing from other beetles.

Dimitry Schidlovsky

Researchers are learning that every dung pat is a complex microcosm unto itself, a teeming habitat not unlike a patch of wetland or the decaying trunk of an old redwood, although in this case the habitat is thankfully short-lived. For scarabs, it may be said that waste makes haste, and entomologists have discovered that as many as 120 different species of dung beetles and tens of thousands of representatives of those species will converge on a single large pat of dung as soon as it is laid, whisking it away within a matter of hours or even minutes.

"If it weren't for dung beetles," said Dr. Bruce E. Gill, a scarab researcher at Agriculture Canada, a government agency in Ottawa, "we'd be up to our eyeballs in you-know-what."

The diversity of beetles that will flock to a lone meadow muffin far exceeds what ecologists would have predicted was likely or even possible, and scientists are being forced to rethink a few pet notions about how animals compete for limited goods and what makes for success or failure in an unstable profession like waste management. They are learning that beetles have evolved a wide assortment of strategies to get as much dung as possible as quickly as possible, to sculpt it and manipulate it for the good of themselves and their offspring, and to keep others from snatching away their valuable booty.

Researchers are also realizing that chance and good fortune play a far greater role than they had thought in determining who reaches a prized resource first and who is able to make the most of it.

The knowledge they are gleaning about the dung beetle community also applies to their understanding of how species compete for more conventional resources, including plants or prey animals.

"I'm fascinated by the enormous diversity of dung beetles that you can see in one dung pat," said Dr. Ilkka Hanski of the University of Helsinki in Finland. "I don't know of any other insect community where such large numbers could be seen in such a small area. It is extraordinary."

Many of the latest findings have been gathered into a new book, *Dung Beetle Ecology,* edited by Dr. Hanski and Dr. Yves Cambefort and published by Princeton University Press. The book is intermittently technical and arcane, but it nevertheless manages to accomplish the seemingly impossible task of transforming a beetle that one previously might have preferred not to dwell on at all into an insect of such worthiness, respectability and even charm that one would like to immediately order a few hundred thousand to help clean up the streets of one's hometown.

Dung beetles, it turns out, are among humanity's greatest benefactors. Not only do they remove dung from sight, smell and inadvertent footstep, but by burying whatever they do not immediately eat they add fertilizing nitrogen to the soil. "Experiments have shown that by burying the dung underground, the beetle increases the amount of nitrogen getting from the dung into the soil, as opposed to being lost in the atmosphere," said Dr. Gill.

Like earthworms, the beetles churn up and aerate the ground, making it more suitable for plant life. Dung beetle larvae consume parasitic worms and maggots that live in dung, thus helping to cut back on microorganisms that spread disease.

"They revitalize the soil, they eliminate noxious wastes we don't like, they keep pastures clean," said Dr. Brett C. Ratcliffe, curator and professor at the University of Nebraska State Museum in Lincoln. "They do so many beneficial things, but if you ask a person on the street what they've heard about dung beetles, they'll look at you like you're crazy."

As beetles go, scarabs are exceptionally sophisticated. In Africa and South America, where some species are the size of apricots, the beetles may couple up like birds to start a family, digging elaborate subterranean nests and provisioning them with dung balls that will serve as food and protection for their young. And these dung balls, called brood balls, are not slapdash little marbles. With a geometric artistry befitting the sculptor Jean Arp, the beetles use their legs and mouthparts to fashion freshly laid dung into huge spherical or pear-shaped objects that may be hundreds of times the girth of their creators. Some beetles even coat the balls with clay, resulting in orbs so large, round and firm they look machine-made.

"When they were first found in expeditions in India and Africa," said Dr. Ratcliffe, "people thought they were cannonballs."

Still working as a duo, the beetles then roll each ball away from the dung pat and down into the underground nest. The female lays a single egg in each brood ball; among the largest species, there may be only one ball and thus one baby per couple.

Safe within its round cocoon, the larva feasts on the fecal matter. As the infant develops over a period of months, the mother stays nearby and tends to the brood balls with exquisite care, cleaning away poisonous molds and fungi and assuring that her young will survive to emerge from its incubator as an adult. That sort of maternal devotion is almost unheard-of among beetles, which normally lay their eggs in a mindless heap and lumber away.

Other scarabs are superspecialists, their habits streamlined to harvest the ordure of one type of mammal alone. Such beetles may cling to the rump fur of, for example, a kangaroo, a wallabee or a sloth, awaiting the moment when the final stage of mammalian digestion is complete and then leap onto the droppings in midair.

Some beetles dine solely on giraffe waste, others on the excretions of wild pigs. Some Panamanian beetles will fly each morning up to the treetop canopies where howler monkeys sleep. They wait for the primates to awake and do their morning business, quickly latching onto the released flotsam and sailing with it 100 feet to the ground, where they then can bury it.

But the majority of dung beetles are generalists rather than specialists, able to make a meal and outfit a nest with any droppings they can find.

"One of the interesting things about the dung beetle family is how some are very species-specific, and some will use any kind of dung they can get their feet on," said Dr. Bernd Heinrich of the University of Vermont in Burlington.

Of keenest interest to the beetles are the generous patties provided by large herbivores, which by the nature of their digestive system must void themselves frequently. The average cow produces 10 to 15 large pats per day. Elephants will provide about four pounds of waste every hour or so, and it is an elephant pat that can become a pulsating Manhattan of beetles, with different species exhibiting a huge variety of tactics. Big scarabs will roll huge balls of it to their nests several yards away, sometimes pushing the balls over logs and boulders; smaller dung-rolling species will shove off with more modest portions. Another class of beetles, called tunnelers, will inter big hunks of dung right beneath the pat, while other, pin-sized beetles will live within the pat itself, munching on it even after it has begun to dry up and be of little use to the larger, more aggressive scarabs. Robber beetles will try to sneak in and pilfer balls painstakingly shaped by others. Joining the fray are many species of dung-eating flies.

"It's like a fast-food outlet, with everybody heading toward it to get a piece of the action," said Dr. Gill. A dung heap is also a sort of singles bar, where beetles in search of mates can meet and begin the joint effort of gathering the goods for their nest. Some of the larger beetles use dung in their courtship dances, the male lifting a deftly rolled bit of dropping and shaking it provocatively in the face of a female.

All of which means that little dung will go to waste. One research team in Africa reported counting 16,000 beetles on a single elephant dung pat; when the scientists returned two hours later, the pat had disappeared.

The incentive to move quickly is great. Not only does every beetle want to get away with the biggest slice of the pie, but while they are scavenging in an exposed heap of dung they are extremely tempting to many insectivores.

"You'll see birds, mongooses, monkeys and other small predators picking around in elephant dung," said Dr. Jan Krikken of the Rijksmuseum of Natural History in Leiden, the Netherlands.

To counteract predation, some beetles have evolved persuasive disguises. One species that frequents elephant dung resembles an undigested stick of the type commonly found in the herbivore's droppings.

Behind the diversity of dung beetles is the resource they live on. Hard though it is to fathom, dung is an exceedingly appealing resource. Most mammals only digest a fraction of the food they eat, and whatever they discard is rich in proteins, nutrients, bacteria, yeast and other sources of nourishment.

Best of all, dung is easy. Animals fight back against would-be predators, and plants generate poisons to ward off herbivores, but dung does not bother defending itself against consumption. "It's available, everyone defecates, and it's the line of least resistance," said Dr. Robert D. Gordon of the department of systematics entomology at the Agricultural Research Service in Washington.

Dung beetles may prefer the droppings of the biggest mammals, but the insects originated more than 350 million years ago, before the appearance of such mammals. Scientists speculate that dung beetles may have fed on dinosaur waste, but no beetle fossil has ever been found in the midst of the petrified dinosaur dung. With the rise and spread of the great mammals around the world, dung beetles likewise began to diversify and multiply. Indeed, the two events occurred in parallel, and some researchers have suggested that large mammals may never have reached the population densities seen in places like the African savanna without the aid of beetles to clean up their waste, thus allowing the plants they feed on to keep growing.

"They're key organisms in the environment," said Dr. Heinrich. "Small beetles may allow a large diversity of herbivores to exist where otherwise a smaller density would have to be."

From the dawn of agriculture and the domestication of waste-heavy livestock animals, human beings also have recognized the incalculable value of the beetles, the ancient Egyptians having taken their admiration to the greatest lengths. Dr. Cambefort of the Natural History Museum in Paris has proposed that the Egyptians' tradition of mummifying their kings and burying them in pyramids was modeled after the burial of a beetle larva in a dung ball. Just as a beetle rises from dirt to a new life, Dr. Cambefort suggests, so the Egyptians believed their pharaohs would be reborn from their interred cocoon. In other words, he said, the Great Pyramids of Giza may be nothing more than glorified dung pats.

The benefits of scarab beetles have not gone unnoticed in our own time. *Dung Beetle Ecology* recounts the ambitious and largely successful effort by the Australian Government to import thousands of exotic dung beetles to help reduce the mountains of dung generated by cattle and sheep. Those livestock animals had themselves been brought to the continent within the past two centuries, and indigenous Australian dung beetles, accustomed to moderate bits of kangaroo and koala dung, were unable to handle the foreign animals' enormous output. By the 1960s, the fecal problem had reached crisis proportions, and the dread native bush flies, which breed in excrement, had reached levels pestilent enough to give birth to the famed "Australian salute," a brush of the hand across the face to swipe away flies.

"If you had a barbecue outside in western Australia, you'd have so many flies on the meat that you wouldn't be able to see it on your plate," said Dr. Bernard M. Doube of the Commonwealth, Scientific and Industrial Research Organization, a government agency in Adelaide. But with the successful introduction of two dozen species of beetles from Asia, Europe and Africa, the dung problem has begun to subside. In the past two or three years, many parts of southern and western Australia have been almost entirely freed of the dung-breeding flies, and pastures that once were coated with a carapace of cattle dung have been restored to useful verdancy.

"What we've done is one of the most ambitious programs for biological control ever undertaken in the world," said Dr. Doube. But he said the program may have been too effective for its own good. "People tend to forget things that aren't there to bother them anymore," he said. "So now they underrate what we've done." The United States Government has also imported beetles from Africa and Asia to help recycle cattle waste in Florida and south Texas.

On a more theoretical scale, ecologists have also learned much from the insects. Scientists historically believed that more than one species could not coexist in the same ecosystem without showing some differences in their use of resources.

"The general rule was that one competitor would eventually prevail over the other," said Dr. Hanski. "There were mathematical equations showing that it must be true."

But given the diversity of dung beetles living on a single resource, nature obviously was not obeying the equations, he said.

Entomologists investigating the beetles have realized that dung as a resource has a few distinguishing characteristics. It is far more ephemeral than, say, a patch of flowers or a burrow of rodents, being here today and gone today. And its distribution in the environment is exceedingly random, with no easily defined rules about when or where it is likely to appear. To most animals, everywhere and anywhere is a potential lavatory.

Therefore, said Dr. Hanski, a strong element of happenstance must be figured into any calculation of the dynamics of the dung beetle community. As it turns out, randomness fosters the survival of many competing species. Some of the larger dung beetles may be inherently better than others at monopolizing prodigious quantities of dung once they get to the pat, he said. But because a smaller, weaker beetle is as likely to be close to the site of miraculous presentation as is a larger and more aggressive beetle, the weaker species will always have a shot at a food source, and the superior scarab will not be able to systematically outcompete it.

"The randomness of the distribution of dung adds a crucial element of chance to survival," he said, "and that element incidentally favors the coexistence of many species."

He said that while dung is an extreme example of an unpredictable resource, other types of ecosystems are likely to be riddled with random fluxes that affect the balance of species and that have yet to be identified. In nature's casino, fortune as well fitness determines survival.

—NATALIE ANGIER, December 1991

The Life Cycle of the Monarch
The female monarch butterfly, at bottom, lays its eggs on the milkweed plant, the egg, insert, hatches into a caterpillar, which feeds on the plant until fully grown. It then spins a silken pad to anchor itself and changes its form into a chrysalis, from which it later emerges as a butterfly.

Patricia J. Wynne

Monarchs' Migration: A Fragile Journey

IN THE FIR-COVERED MOUNTAINS of southern Mexico, hundreds of millions of monarch butterflies packed in tight, brilliant clusters are now settling in for their winter rest after completing one of nature's most extraordinary feats. Each year, the insects migrate as far as 2,500 miles between their summer breeding grounds in the northern United States and Canada and their winter retreats in Mexico.

This splendid natural phenomenon can no longer be taken for granted. The butterfly's special wintering sites in 13 Mexican mountain enclaves—and in certain "monarch groves" in coastal California, where a smaller, separate cohort spends the winter—are threatened by logging and development. Vigorous conservation efforts in Mexico may have helped secure the monarch's refuges there. And voters in Pacific Grove, California, approved the $2 million purchase of a privately owned monarch grove to save it from development.

But the battle is far from won, scientists and conservationists say. "With a small amount of human negligence, everything could disappear," said Carlos Gottfried, the chairman of the board of Monarca, a nonprofit organization based in Mexico City that has been spearheading a vigorous effort to preserve the Mexican wintering grounds.

And Dr. Lincoln P. Brower of the University of Florida, an expert on the monarch, fears that despite the Mexican conservation campaign, "we could lose the whole migration to Mexico in the next decade or so."

What makes the monarchs' migration so special is that the butterflies successfully navigate their path to wintering grounds they have never seen: The butterflies that leave the Mexican winter retreats to head back northward in the spring are the great-grandparents of those that return in the fall.

Catching rides from Canada to Mexico on winds and spiraling columns of warm air, these expert little gliders in orange and black finery set their course unerringly toward faraway destinations. Birds routinely migrate such long distances, but no other insects are known to do so. The returning monarchs, each born in the north, rely solely on navigational instructions programmed genetically into one of the tiniest of nervous systems.

"If you've ever looked inside the brain of a butterfly, it's about the size of a pinhead," said Dr. Brower, "and yet the minicomputer inside that pinhead has all the necessary information to get them to Mexico without having been there before."

How this guidance system works is a mystery, and the prospect of someday understanding such a "complex neuronal control system" is reason enough in itself to preserve the monarch migrations, Dr. Brower said.

Scientists do know that the annual flight of the monarchs is part of an ecological relationship among the butterflies, their habitat and the climate that is as fragile as the tissue-winged insects themselves. Conservationists wish to preserve the whole ecological framework because it is what makes possible the natural wonder of the migration. The monarch would disappear from almost all of North America if the migration ceased, although nonmigrating populations would continue to exist in southern Florida and parts of the tropics.

Fortunately for the conservation effort, the monarch migration is beginning to acquire a mystique akin to that of the great whales and the African elephant. Growing numbers of tourists flock to marvel at the quivering masses of monarchs that festoon the trees in the wintering areas.

Residents of those areas invest the monarchs with a pride that sometimes borders on reverence. In Pacific Grove, they are the biggest thing in town. Motels are named for them. Children dressed in monarch costumes parade through the town each fall, when the butterflies appear.

In Mexico, the insects' arrival at the beginning of November coincides with a religious observance in which the butterflies, according to a mythology going back to pre-Columbian days, are seen as the returning souls of the dead. And in the United States, the monarch is a front-runner, along with the honeybee, in a continuing campaign to name a national insect.

The monarch's glamour, in the view of some conservationists, makes it an ideal test of the willingness of North Americans to care for an ecological treasure.

"If the people can and will save their monarchs, perhaps they will be ready to think about other beneficial insects; only then will we see a true popular campaign for biological diversity," Robert Michael Pyle wrote last year in the journal of the Xerces Society, an international organization dedicated to the preservation of invertebrate habitats. The society has been active in the campaign to preserve monarch habitats in California.

While the conservation effort goes on, a small band of scientists continues to tease out the details of the migration itself.

The origins of the migration lie in the monarch's dependence on the milkweed plant and in the inability of the insect, as a tropical creature, to withstand cold weather. In ancient times, scientists believe, the milkweed plant moved up from its home in the tropics to colonize North America and the monarchs followed their food plant northward.

Milkweed contains bitter poisons, called cardiac glycosides, that originally evolved as a defense mechanism to protect the plants from insect predators. The monarchs not only adapted to defeat the poison but converted it to a chemical weapon for their own defense. The monarch's caterpillars feed on the plants, storing the toxic chemicals in their bodies as they grow. The adult butterflies retain the poisons, which make predators throw up. A bird that has tasted one monarch never tries another.

Scientists believe that the distinctive orange and black markings of the insect serve as a warning signal that birds observe. Another North American butterfly, the viceroy, has evolved markings similar to the monarch's, and this mimicry helps protect it even though it lacks the monarch's chemical defenses.

Two species of birds have adapted to the poison and feast on the butterflies. So has a mouse that gorges on the somnolent bodies of monarchs passing the winter in Mexico. So far these predators have not significantly reduced the monarch's numbers.

The United States enjoys two different populations of monarchs, one to the east of the Rockies and one to the west. The western monarchs spend the winter in the groves of California. Monarchs east of the Rockies are the offspring of butterflies that overwintered in Mexico.

The winter refuge in Mexico, discovered in 1974, consists of 13 compact wintering sites scattered in a small 75-by-35-mile area in the mountains 75 miles west of Mexico City.

The sites are ideal, said Dr. Brower, because they maintain a climate that is just right for the monarchs at a time when their highest priority, aside from survival, is to conserve energy for the return migration in the spring. The wintering area is just south of the Tropic of Cancer and its temperature is relatively stable. The butterflies roost in mountainside fir trees within a narrow altitude band ranging from 9,500 to 11,000 feet, their gaudy bodies sometimes festooning a tree so thickly that neither branch nor needle can be seen.

At that altitude, the air is warm enough to keep the butterflies from freezing but cool enough so that they do not burn up calories unnecessarily. They fly around when warmed, wasting energy. The high mountains also capture moisture that bathes the butterflies and prevents water loss.

A similarly suitable microclimate exists on the California coast where the wintering sites of the western monarchs are situated.

At the end of their autumn flight from colder climes, the monarchs arrive in Mexico robust and unmated, their brilliance as fresh as if they had just emerged from the chrysalis. They are superbutterflies with a nine-month life span, living longer than any others. In the spring, they awaken from their winter dormancy as feverish, single-minded lovers, rushing pell-mell to where the first milkweeds are coming up along the United States Gulf Coast.

"We don't know too much about their flight north, but we do know one thing—they're in more of a hurry" than on the fall return flight, said Dr. David Gibo, a biologist at the University of Toronto who has studied the monarchs' flight habits. "It's a race to the milkweed."

Because of their haste, he said, the butterflies appear to use up much of their energy in flight so high-powered that although millions leave Mexico, relatively few reach the United States. Those that do arrive come in low, almost on the deck, males searching for females and both sexes searching for milkweeds.

"A male hangs around these milkweed patches, and if a female comes through, he'll just go after her like a pursuit ship," said Dr. Brower. "If there are lots of males around, they'll harass the female, who has lots of ways of evading them. She might fly through the branches of a tree, and the male gets lost. Then she goes and lays her eggs."

This furious expenditure of energy drains the parent butterflies of life. Their offspring fly off northward, following the milkweed as it appears, and by summer have dispersed across the northern half of the United States east of the Rockies, ranging as far north as North Dakota, southern Ontario and Maine.

The western migration is smaller and less dramatic. In the spring, the butterflies leave their refuges on the California coast. Their first new generation is born on the slopes of the Sierras, and subsequent movement takes the monarchs into Idaho, Nevada, Utah and as far south as Phoenix.

Dr. Brower and colleagues in Florida just last month confirmed that the first generation born in the East appears in the Gulf states rather than farther north. They did so by analyzing the kinds of poisons contained in butterflies' bodies. Some species of milkweed grow only in certain parts of the country, and the poisons vary in a way that provides a "fingerprint" by which it is possible to identify which species of milkweed nourished the butterfly—and therefore, in what part of the country the butterfly was born.

East of the Rockies, three or four more generations of butterflies are born after the group born on the Gulf Coast flies to its northern range. Each of these generations lives for about three weeks, except the one tapped by nature to complete the cycle by heading south once again.

In August, the reproductive organs of this generation become dormant. Its members lose interest in sex, but they become very gregarious and irresistibly attracted to flowers and each other. They cloak goldenrods, daisies and other composite flowers in what Dr. Brower calls "social drinking groups," sucking up nectar to nourish them on their journey.

In contrast to the urgent flight of their great-grandparents from Mexico in the spring, Dr. Gibo said, these southward-flying monarchs take it relatively easy. He has studied their flight tactics, tracking them from his glider and with ground radar.

He has found that the butterflies expertly exploit ascending spirals of warm air, called thermals, then glide downwind until hitching onto another thermal. Glider pilots, Dr. Gibo said, have seen monarchs circling as high as three quarters of a mile.

If a crosswind interferes with their course, the monarchs can somehow correct for it, to keep on course southwestward to Mexico.

Faced with a headwind, the monarchs simply drop to Earth and look for nectar. When they arrive in southern Texas and northern Mexico, they go on a feeding binge designed to see them through the winter. "By the time they get to the overwintering site," Dr. Brower said, "they're literally butterballs."

Conservationists concerned about threats to the butterflies' overwintering places are encouraged by last month's vote in Pacific Grove, California, to protect the town's monarch site. "I'm really proud of them," Dr. Brower said. "They had to get sixty-seven percent of the vote and they got sixty-nine percent."

Still, the butterflies remain in a precarious position in both California and Mexico, said Curtis Freese, vice president for regional programs of the World Wildlife Fund, which is aiding the Mexican conservation project. With just a few small wintering sites serving the entire population, he said, "one can never feel entirely easy, can never rest."

Dr. Brower expressed doubt that the Mexican preservation efforts would be wholly successful. In 1986, the Mexican Government banned logging in the overwintering areas. But since then, Dr. Brower said, he has seen loggers cutting trees "right in the butterfly colonies." He said loggers had also cut empty trees used by monarchs in previous years, and to which they might have returned in the future.

Mexico has turned one of its 13 wintering sites into a tourist attraction, and plans to convert a second, in hope that local people will make money off the tourist trade and perceive protection of the roosting trees to be in their own interest. Some 70,000 tourists visited the first site last year, said Mr. Gottfried of Monarca, the Mexican conservation organization, and the number is steadily growing.

Mr. Gottfried, who has been fighting the battle to preserve the monarch migration for 13 years, said he believed that the Mexican Government and public were giving the monarch issue serious attention.

Mr. Gottfried said that during a recent field trip to the 13 sites, "I was very pleased with what I saw."

"There's been no cutting this year in the core area," he said. "This is the first year I can say that with total conviction."

Dr. Brower responded: "I hope they've stopped the cutting, but I strongly doubt they have. Sometimes the loggers don't get going till the height of the dry season in February.

"The Mexicans seem to respect the butterflies," he said. But until it is clear that cutting has stopped, he said, there is danger of "a catastrophe that's going to spell the end of monarch butterflies in eastern North America."

—WILLIAM K. STEVENS, December 1990

Can You Like a Roach?
You Might Be Surprised

IF ABSENCE MAKES THE HEART GROW FONDER, then perhaps the moment has arrived to consider a modest celebration of the cockroach.

In recent times, many city dwellers have been able to stride into their kitchens at night with a newfound confidence that they can flick on the light, take a glass from the cupboard, even grab a few cookies from a box on the counter—all without the odious sight of dozens of greasy brown cockroaches skittering for cover.

A new generation of insecticides, packed into discreet little disk-shaped bait traps called Combat or applied in more potent concentrations by professional exterminators, has helped bring the ubiquitous German cockroach to its six spindly knees.

The creature is far, far from nearing extinction, and indeed remains a serious pest in restaurants, hospitals and many inner-city housing projects. But entomologists and public health officials said that since the new insecticides, amidinohydrazones, were introduced in the mid-1980s, they have made a significant dent in the less extreme cases of infestations.

"Almost everyone I've talked to, both personally and on the job, has noticed a vast change in the roach population," said Roz Post, a spokeswoman for Housing Preservation and Development in New York. "There was a time when people were horrified at roaches running rampant, and now everybody keeps saying, 'Where did they go to?'"

Entomologists report that the new chemicals will cut the German cockroach population by 50 percent to nearly 100 percent, depending on the severity of the infestation. More heartening still, the latest studies of cockroaches collected from around the country indicate that the insect is showing no signs of developing resistance to the amidinohydrazones, as it has to nearly every other noxious compound leveled against it in the past.

"I've gathered up populations from a dozen or so geographical locations," said Dr. Donald G. Cochran, a cockroach and insecticide expert at Virginia Polytechnic Institute and State University in Blacksburg. "I haven't seen any indication of resistance, and I don't think you're going to find any."

And should the creature somehow manage to mutate beyond the might of the current pesticides, other new and highly effective compounds are being tested, many of them based on subtle understanding of the insect's biology and habits.

"We have some excellent materials coming up," said Dr. Austin Frishman, an entomologist and pest control consultant in New York's Farmingdale, Long Island, who travels around the world to help businesses suffering from cockroach infestations. "The chemistry is there to keep roaches under control for the next ten years if we play our cards right."

So, now that humans no longer need share every meal and inch of shelf space with unwelcome squatters, entomologists hope they can instill, if not outright affection, at least a detached sense of admiration for cockroaches, which are among the oldest and most resourceful of all land animals.

In new studies of species found in the tropics—where the creatures know their place, and that place is not ours—researchers have discovered that the insects display a wide range of impressive behaviors. "Cockroaches do quite a few things that we normally associate more with mammals than with insects," said Dr. Coby Schal, an entomologist at the Cook College of Rutgers University in New Brunswick, New Jersey, who has studied cockroaches in Costa Rica and other Central American countries.

Some female cockroaches are devoted mothers, carrying their offspring in little pouches, kangaroo-style, rather than simply dropping their eggs and leaving them to their own devices, as many insects do.

One scientist has recently discovered a type of cockroach that does the insect equivalent of breastfeeding.

In many animal species, the male's only contribution to the progeny's welfare is the donation of his genes. The male cockroach carries paternal care to a rather greater length. He will dine off bird droppings for the sole purpose of extracting precious nitrogen that he can bestow on his developing offspring.

One kind of cockroach that lives in Central American tree bark turns out to be as social an insect as termites or bees. The males and females pair off to nurture their immature forms, known as nymphs, for the five or six years it takes the species to reach adulthood. All members of a nest maintain a sense of group identity and cooperation through the use of mutual grooming, antennae stroking and placating pheromones, chemical signals that are secreted by glands on the thorax of one insect and detected by the antennae of another.

Cockroaches are exquisitely sensitive to the slightest breezes, a trait that accounts for their unusually long antennae. Such tactile sensitivity, combined with a nervous system built of exceptionally large cells, makes the cockroach an ideal experimental organism for the study of how nerve cells work.

"Among neurobiologists, the cockroach has become the insect version of the white rat," said Dr. May R. Berenbaum, an entomologist at the University of Illinois at Urbana-Champaign.

Dr. Ivan Huber of Fairleigh Dickinson University in Madison, New Jersey, author of a book published last March by CRC Press called *Cockroaches as Models for Neurobiology*, said that cockroaches are "beautifully easy" to study. Their receptors, which detect a single chemical molecule or a puff of air, are on the outside of the body, where they can be readily manipulated. And the cockroach's head will live and respond for at least 12 hours after the animal has been decapitated. Furthermore, few animal-rights activists will disrupt a laboratory where the experimental organism is a cockroach.

"People get all worked up over using kittens and puppies for medical experiments," said Dr. Berenbaum. "But nobody is going to shed any tears if you kill a few thousand cockroaches for the good of science."

Dr. Berenbaum has made it her particular mission to bolster the cockroach's reputation. Every year she holds an "insect fear film festival," using clips from movies to tweak the public's interest in insects and to dispel myths. The theme of the 1991 festival, presented last month, was the cockroach.

"There are a surprising number of cockroach-oriented films," she said, "probably because it's easy to breed them in large quantities."

In many of the shorter films, as well as in animated movies, she said, the cockroach is a sympathetic character. One offering, *All's Quiet in Sparkle*

City, an antiwar film from the early 1970s, equates efforts to eradicate cockroaches with genocide. A 1989 comedy, *Dr. Ded Bug,* is shot from the insect's perspective, as a frenzied chef attempts to hunt down and kill a cockroach. Cartoon cockroaches talk in high, chipper voices and rarely stop smiling. "They're modeled after Mickey Mouse," she said. "In animated films, vermin become your friends."

Whether or not cockroaches become one's bosom buddies, Dr. Berenbaum and other entomologists say the insects merit respect for their antiquity and their diversity. Fossils of cockroach-like species have been found dating to 280 million years ago, and some entomologists estimate that the creatures may be as old as 400 million years. By contrast, beetles are only about 150 million years old, while butterflies are a youthful 60 million years old.

Cockroaches are found in nearly every part of the world, but the great majority of the 4,000 known species live in the equatorial belt, and entomologists believe another 6,000 tropical species remain to be discovered. Cockroaches range in size from a quarter of an inch to the forbidding Megablatta of Central America, which in length and girth approaches the dimensions of a small rat. Universal to cockroaches are long, segmented antennae; a leathery pair of front wings that allow many warm-weather species to fly but otherwise are vestigial; and the famous cockroach head, which is tucked under and slightly pointed toward the rear.

Some smaller cockroaches are exquisitely colored, and may be deep crimson, spring green, a creamy white or a pale toffee. The most dazzling specimen is a cobalt blue with bronze flecks and thin red stripes.

"I thought it was a beetle when I first saw it," said Dr. William J. Bell of the University of Kansas in Lawrence, who has studied many tropical cockroach species. "If it were any bigger, people might try to put it in a birdcage."

But only one type of cockroach is frequently kept as a pet: the three-inch Madagascar hissing roach, which attempts to scare off predators by expelling a noisy blast of air through holes in its upper thorax. The hissing roach is covered with an armor-like cuticle that makes it far more appealing to hold and stroke than many cockroach species. The German cockroach is coated with an oil that eases its passage into cracks thinner than a paper match.

The most advanced of all cockroaches, in terms of evolution's tree, is a species known as *Diploptera punctata,* according to Dr. Barbara Stay, a biologist at the University of Iowa in Iowa City. The female carries her embryos live, rather than in an egg case, and she is the only insect known to nourish her young in the uterus. The lining of the brood pouch, where about 12 baby cockroaches grow at a time, secretes a substance that Dr. Stay calls cockroach milk. Like mammalian milk, it is rich in protein, carbohydrates and fat.

"The cockroach milk is not produced until the embryos have a fully developed digestive tract," said Dr. Stay. "Then they sit in the pouch drinking up the milk right through their mouths."

But while some cockroaches have evolved an elaborate system of maternal care, others have opted for greater fecundity and the utmost behavioral flexibility, and these are the species that have become pests to humans. Only 20 types of cockroaches are classified as pests, and only two of these, the German and the American cockroaches, are broadly familiar, the three-inch American cockroach going by the name of palmetto bug in Florida and the water bug in New York. And these two pest species have fared so successfully in their strategy of taking up residence with humans that they no longer have an independent existence or any representatives of their kind in nature.

"We've looked everywhere for their natural habitat," said Dr. Bell. "But every time we thought we had seen one in the wild, there's turned out to be somebody's house nearby."

The smaller German cockroach is an especially prolific breeder, able to spawn 30 or 40 infant cockroaches every three weeks. If the population were left unchecked, a single female German cockroach could theoretically give rise to about 40 million offspring in her two-year life span.

The insects grow to adulthood rapidly and molt frequently, which is why cockroaches can present a real health hazard for those with allergies. Dr. Richard J. Brenner, a research entomologist with the Agricultural Research Service and Veterinary Entomology Research Laboratory in Gainesville, Florida, estimates that as many as 15 million Americans suffer from cockroach allergies, as their immune system mounts an overzealous defense against airborne particles of molted cockroach skin.

The allergies often worsen with time and continued exposure to cockroaches, and entomologists who work with the creatures said they, too, developed wheeziness, skin rashes and sinus troubles after years of pursuing their research.

"I love cockroaches, and I don't know where I'd be without them," said Dr. Schal. "But unfortunately I'm starting to get slightly allergic to them."

For that reason, and because of the possibility that cockroaches may transmit the unsavory microbes piggybacked upon them, even entomologists who like the insects in the wild spend part of their time devising better ways to thwart cockroach pests.

To determine where cockroaches congregate and why, Dr. Brenner and his colleagues have designed an entire mock house. Its 200 sensors monitor the microclimate every 65 milliseconds and at every possible spot—behind walls, under the sink, up in the rafters. The researchers have learned that nothing will repel cockroaches as surely as ventilation. The animals use almost indetectable air currents to sense the chemical signals of their mates, but any air movement approaching a draft will quickly and fatally desiccate the cockroach's coating. Dr. Brenner believes that by designing homes to have air circulating throughout, even in cupboards, the cockroach problem can be largely curtailed.

Others are encouraged by the new generation of pesticides, which differ markedly from the older poisons, often used in the spray cans. Those pesticides, which include the organophosphates and carbamates, are potent nerve poisons that block the transmission of impulses from one nerve cell to the next. But because the older poisons work on only a single component of the nervous system, some cockroaches have turned out to have an inborn resistance to that one means of attack. And those insects have been the ones that have survived to propagate entire legions of resistant nymphs.

By comparison, the newer pesticides seem to be more global in their activity, affecting so many parts of cockroach physiology that it is unlikely any one insect will have all the genetic traits necessary to withstand the assault. The active ingredient found in Combat, for example, interferes with multiple steps in the biochemical pathway that allows a cell to use its stores of energy.

Of perhaps greater importance, the toxin works at extremely low concentrations. Dr. Cochran believes that one reason cockroaches may have trouble developing resistance to the chemical is that even those few insects able to survive a nibble of poisoned bait become sterilized. Thus, they fail to pass along their detoxifying ability to offspring, as survivors of the older generation of pesticides were able to do.

But entomologists realize that, over the long term, the task of keeping cockroaches at bay is formidable.

"The insect has all the biological factors that help it survive," said Dr. Michael K. Rust, an entomologist at the University of California at Riverside. "It has a high reproductive rate and a fast life cycle. It has been living with man for thousands of years. So anytime I hear about a new chemical or bait I ask myself, How long will this last?" Not long enough, surely, for people to start missing their nocturnal companions.

—NATALIE ANGIER, March 1991

Dragonflies' Perfection
as Performers and Predators

IF THERE'S ANYTHING that can make a grown-up feel silly and inept, it's a dragonfly. Just try to catch one, even with a net, and you'll soon discover the 300-million-year-old advantage of having evolved with 360-degree vision and the ability to weave, dodge, hover, fly backward and change directions in a fraction of a second, then take off at 35 miles an hour.

Studying dragonflies, as their small legion of enthusiasts is quick to admit, can require extreme patience and a willingness to be repeatedly humiliated by a featherweight insect that seems to take pleasure in tormenting its would-be captors, coming ever so close, only to dart away as an elongated butterfly net swoops futilely through the air. Where the would-be captor's muscles work in unison, each of the insect's four wings operates individually under direct muscular control, giving it extraordinary maneuverability.

Patience does occasionally have its rewards, however. William Smith, a zoologist with the Wisconsin Department of Natural Resources Natural Heritage Inventory Program, is about to report his second discovery of an unknown species of snaketail, a family of dragonflies whose adult bodies resemble a snake in reverse, with a reptilian head at the insect's tail end.

Assisted by Timothy Vogt, now at the Illinois State Museum in Springfield, in 1993 Mr. Smith published the description of his first newly discovered species, the St. Croix snaketail, named for the scenic St. Croix River between Wisconsin and Minnesota where it breeds and feeds. Now he, Mr. Vogt and Kenneth Tennesson of the Tennessee Valley Authority are preparing to describe another new species found along the Eau Claire River in western Wisconsin.

Fighter Jets That Can Also Hover and Dart

A newly identified species of dragonfly, the St. Croix snake-tail, has a voracious appetite for mosquitos and blackflies. The highly maneuverable wings are under individual muscular control. They change shape in response to aerodynamic forces, reducing strain on the powerful flight muscles in the thorax.

A Complicated Sex Life

Dragonflies are often observed in the acrobatic feat of mating on the wing. Males have two reproductive organs, one at the rear end that produces sperm and one farther forward on the underside, where sperm is transferred before mating. Clasped by appendages at the male's abdominal tip, the female brings her abdominal tip into contact with the male's copulatory organ and receives the sperm.

A Fatal Traffic Jam

When the adult emerges from the nymphal case, its body and wings stay soft for hours until they harden in the air. In certain lakes, a number of species may emerge at the same time, and soft-bodied adults can be trampled and fatally punctured by later waves of hard-bodied nymphs.

Special Structures for Feeding

In their immature nymph form, dragonflies are also voracious feeders. Most have a lower lip, or labium, that can be extended and retracted at high speed to help them capture prey. The labium brings the captured insect directly into the mouth, which is flanked with slicing mandibles.

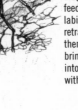

Michael Rothman

"In the continental United States, finding a new species is not an everyday occurrence," Mr. Smith said with modest pride. "Most workers in the field—there are maybe two in each state—have not described a new species. But in the last few years there's been an upsurge in interest in dragonflies, and when you begin to look very carefully, new species show up."

Though dragonflies have inhabited Earth since the Carboniferous era, predating birds by millions of years, they are just now beginning to attract widespread interest among entomologists.

Fueled in part by the potential for finding new species, "popular interest has blossomed in recent years and begun to rival the attention once paid to butterflies," Mr. Smith said. "There is so much yet to learn about how and where dragonflies live."

As one of the first organisms to fly, dragonflies are clearly among the oldest and most successful of insects. Giant dragonflies with the wingspan of a hawk were dinosaurs' contemporaries. Dr. Donald G. Huggins, an aquatic ecologist at the University of Kansas in Lawrence, said: "They are as close to being the perfect predator as anything on Earth. When you're that good, there's no pressure to change."

But like most other modern-day flora and fauna, they are not immune to the many contemporary forces that can render a species extinct.

"As with many other species, the primary issue is habitat destruction," Dr. Huggins said. "People overuse water, which dries up or greatly diminishes the flow of the springs that are the sources of permanent water dragonflies depend on." People are far more conscious of the environment these days, but when it comes to a choice between watering a golf course and preserving dragonfly habitats, the golf course usually wins. "Dragonflies don't have good lawyers," Dr. Huggins lamented.

Although the colorful, flighty, mosquito-gobbling adults are what attract most researchers and amateur collectors, the airborne form lives only a few weeks or months. Most of a dragonfly's life—from one to four years, depending on the species—is spent crawling or swimming underwater as a predatory larva. The growing larva splits and sheds its skin 7 to 14 times as it grows before climbing out of the water to shed one last time and, in a seemingly magical transformation, emerges moments later as a mature dragonfly. Insects farther along the evolutionary chain go from egg to larva and then to a pupal stage before metamorphosing into an adult.

John Haarstadt, an entomologist and ecologist at the University of Minnesota's Cedar Creek Natural History Area, said that the larvae, or nymphs, are "no less interesting" than the more obvious and colorful adults. To escape predators and capture prey, he pointed out, they often rely on water-fueled "jet propulsion" to flit rapidly out of harm's way and on an extendable lower jaw "armed with wicked teeth to capture prey."

Mr. Haarstadt, who surveyed the stream-dwelling dragonflies of eastern Minnesota for the state's Department of Natural Resources, said more attention should be paid to these "valuable indicators of stream deterioration."

But environmental degradation is not the only enemy. The larvae feed on tiny fish, but bigger fish adore the larvae (a fact well-known to fly fishermen, who use many nymph-like flies to reap their sport's rewards), and birds and other predatory insects dine greedily on the newly emerged adults as they sit, highly vulnerable, drying their wings. But even birds have trouble catching adults once they become airborne.

Sometimes dragonflies are their own worst enemy. While studying the coexistence of dragonfly species in one of Minnesota's 10,000 lakes, Mr. Haarstadt noted an "enormous emergence of three different species in late May, with seventy-five percent of the total population of millions of dragonflies emerging within three days." As larvae and adults, the three species occupied different niches and stayed out of each other's way. But when the length of the day and the temperature of the water were just right, the last larval stages of all the three species were programmed to leave the water at the same time.

"The emerging nymphs crawled all over each other and trampled themselves to death. It was the largest cause of mortality," Mr. Haarstadt said in an interview here. But year after year, the same three species continue to coexist and apparently thrive in the lake. Mr. Haarstadt attributes their success to the fact that dragonfly adults "are superb feeding machines that can easily mature several batches of eggs each season." In some species, each adult is capable of eating an estimated 300 mosquitoes a day. Hence the dragonfly's popular nickname, "mosquito hawk," nature's answer to DDT. But the insect-eating adults live only for a few weeks before they mate, lay their eggs and die.

Most dragonflies and their companions in the order Odonata, the damselflies, have common names descriptive of their appearance or behavior.

Some sound like the characters in a B movie—black dragon, blue pirate, green darner, corporal. Others hint of romance—bog dancer, green damsel and amberwing. Still others reflect their appearance—spiny-legged clubtail, raggedy skimmer and variegated damsel. Dragonflies are the ones that alight with wings outspread, while damselflies fold their wings above their bodies. Neither can hold their wings against their bodies like other insects.

Some 450 species of dragonflies and damselflies are known to inhabit North America, and more than 3,500 species flit about worldwide.

Mr. Smith found the St. Croix snaketail while trying to develop an ecologically sound method of determining the presence of dragonfly species. Instead of collecting adults of possibly rare species, he was surveying the mold-like skins cast aside on dry land when the last larval stage leaves the water. While collecting the discarded casts, or exuviae, on the Wisconsin side of the St. Croix during the summer of 1989, Mr. Smith stumbled upon 16 specimens he had never seen before. When he analyzed the summer's collection that fall, he determined from the wingpads on the larval skin and elongated oval abdomen that the insect was a snaketail. But, he said, "the exuvia didn't fit the description of any known species from North America."

Mr. Vogt, then a co-worker in the Wisconsin heritage program, checked descriptions of Asian and South American snaketail larvae. None matched. But to be sure this was indeed a new species, Mr. Smith needed an adult.

In addition to the usual difficulties of snagging an adult dragonfly, there was the fact that "you've got to be there at just the right time of year, on the right day and in good weather." Dragonflies, cold-blooded like all insects, do not fly in the cold and rain. They, like sunbathers of every two- and four-legged variety, like it warm and sunny. Now it was mid-fall with portents of an upper Midwest winter in the air. Instead of waiting until the next summer to try to capture an adult, Mr. Smith decided to look for some larvae lying dormant in the now icy river.

So in November, he recalled, while wading in the frigid river "in water deep enough for me to barely stand in, which is unusual for dragonfly larvae, I came up with half a dozen dormant larvae" that looked just like the skins he'd collected the summer before. Back in Madison, Mr. Smith placed the larvae in a large cold aquarium and gradually warmed the water and increased the exposure to light to break the larval dormancy. Finally, in February, the larvae climbed out of the water, shed their final skin and hatched

into magnificent clear-winged adults with bulging, bright green eyes, emerald thorax and black abdomen with bright yellow markings.

"Right away I knew it was a new species," he said. But to be absolutely
sure he could describe it fully and accurately, he returned to the river the
next summer to snag some of the naturally emerging adults, which he and
Mr. Vogt named *Ophiogomphus* (the genus for snaketail) *susbehcha* (the
Lakota Sioux word for dragonfly), now known as the St. Croix snaketail.

The subject of Mr. Vogt's current study, Hine's emerald dragonfly, a
green-eyed skimmer, is an endangered species whose limited environmental range may dictate its own demise. There are only two known populations living on what Mr. Vogt called "deerfly and mosquito ranches"—one
in the Des Plaines River valley in northeastern Illinois and the other in Door
County in northeastern Wisconsin. He said the insect is restricted to "a very
specialized larval habitat" that he is still trying to define fully. While most
other dragonflies, during their lengthy larval stages, live in ponds, streams,
lakes and marshes, which are plentiful, the Hine's emerald insists on
immersing itself in shallow, lime-rich wetland muck with a bedrock of Nia-
garan dolomite less than 20 inches down. There just is not much of that
kind of terrain around, Mr. Vogt said.

Although some species, like the Hine's emerald dragonfly, are endangered, dragonflies and damselflies are clearly survivors that in many ways
are ideally suited to withstand human incursions. Dr. Huggins said that in
general, odonate larvae are able to resist the toxic effects of most pesticides
and heavy metals. They can also tolerate extremes of acidity, alkalinity, hard-
ness and even low levels of oxygen in the water, an ability he attributed to
a variety of respiratory structures in the larvae. These include a gill-like
apparatus in the tail that extracts oxygen from water. Mr. Haarstadt said the
larvae also use this anal apparatus to escape predators by expelling an aque-
ous jet stream that "propels them through the water like a torpedo."

Mating in dragonflies is itself an unusual acrobatic act. Dragonflies are
often seen flying about in copulatory bliss, hooked together front to rear in
the shape of a heart or circle. Male dragonflies and damselflies have two
reproductive organs, one at the hind end that produces sperm and a second
one on the underside near the front of the body that copulates with the
female. Before it can mate, the male must transfer its sperm to the anterior
organ, a feat accomplished by bending its tail end forward. The male then

uses appendages at his abdominal tip to clasp the female by the thorax or back of the head. Now it is the female's turn for gymnastics. She forms a circle of love by bringing her abdominal tip into contact with the male's copulatory organ to receive the sperm.

To assure his paternity, the male may hang on to the female while she deposits her eggs either by inserting them into underwater plant stems or simply smearing them on the water's surface. In some species, the male lets go of the female while she lays her eggs, but he hangs around at the ready to chase off any competition. And sometimes, even before mating, the male may scrape out the female's genitalia to remove sperm from a prior mating.

More than 300 eggs may be laid at one time, Mr. Haarstadt said, and sometimes several hundred females will lay their eggs in the same spot, producing large egg masses up to two feet in diameter that are easily mistaken for frog eggs. "Why do they do that? Is it for protection? Might the innermost eggs suffocate?" That is just one of the many challenging questions yet to be answered about dragonflies.

—JANE E. BRODY, August 1997

A Student of Bees Decodes the Signal to Swarm

PARTLY SHADED by the slender branches of a paloverde tree, his face just inches from a cluster of 5,000 honeybees, Dr. P. Kirk Visscher spoke into a walkie-talkie. "I think these bees are going to take off," he told a colleague stationed 160 yards west, at a wooden nest box that the bees had been scouting in their search for a new home.

Indeed, the cluster was becoming jumpier by the minute. Worker bees had begun "buzz running," plowing furiously through the masses of their sisters, as if prodding them to get ready. The buzzing grew louder as the entire cluster pulsated and changed shape.

"They're gonna go!" Dr. Visscher cried. His partner, Dr. Scott Camazine, came running to watch.

By dozens and then hundreds, the bees lifted off into the warm light of a February afternoon in the desert 150 miles east of Los Angeles. Within a minute, all 5,000 were airborne, zooming around in a circular holding pattern 20 or 30 feet in diameter, just barely above the heads of the two entomologists and Richard Vetter, a staff research associate from Dr. Visscher's laboratory at the University of California at Riverside. "Streaker" bees that had visited the nest box roared westward across the circle, signaling the others to head that way. A hum filled the dry air.

It was like standing in the midst of a tornado of bees. A visitor's first impulse might be to run for dear life or roll up in a ball on the ground, but it would surely give way to wonder at an event that even textbooks and scientific reports cannot resist describing as "spectacular."

"Very few people ever see this," Dr. Camazine said, smiling.

Such behavior, known as swarming, occurs when part of a bee colony leaves the nest to find a new home, and it has been described in detail by various scientists over the years. But important elements remain unexplained, and they are on a seemingly endless list of things that Dr. Visscher hopes to learn about bees.

An associate professor of entomology at Riverside, Dr. Visscher is both a honeybee biologist and a lifelong beekeeper. Of the 20,000 known species of bees, only about six make honey, and one of them, *Apis mellifera,* the best known and most common, is the focus of much of Dr. Visscher's research. He keeps 60 hives—about three million bees—on campus.

Bees generally swarm in spring, when the hive becomes too crowded. Workers begin rearing a new queen, and the old queen, often against her will, is swept out of the hive by a swarm that may contain 10,000 to 20,000 bees, between 30 percent and 70 percent of the original colony.

The swarm, clustered around the old queen, settles for a few days in an interim spot, often a tree branch, while scouts, arising from the ranks of worker bees, fly off to find a new site for a hive. When they return, the scouts do a dance for their nestmates that is similar to the one forager bees perform to direct others to food. Scouts make up perhaps 5 percent of the swarm. Some visit more than one site, and different scouts dance for different sites. Nonetheless, the moment arrives when the entire swarm takes to the air, and, somehow, all the bees go in the right direction to the new site.

"How do the bees achieve a unanimous decision?" Dr. Visscher asked.

Dr. Camazine said, "It's one of the last remaining great questions about honeybees, and Kirk is the ideal person to collaborate with." Dr. Camazine, an emergency room doctor, has put medicine aside to become an assistant professor of entomology at Pennsylvania State University.

Most researchers think that the scouts visit various sites, compare them somehow in whatever is the bee equivalent of a mind, pick the best one and then dance for it, so that much of the dancing eventually represents one site, which the swarm then occupies.

Dr. Visscher and Dr. Camazine are not so sure that bees make comparisons, and the experiments they have been conducting in Cactus City have been designed to test the idea. They take a small swarm from one of Dr. Visscher's hives and set it on a plywood stand in a stretch of desert that has virtually no sites, such as hollow trees or caves, that would attract bees looking for a home. That way, the swarm will go to nest boxes set up by the researchers, who post themselves at the boxes and mark scouts with a shade of paint that will reveal which box they visited.

Any bee that shows up at one box after visiting another—as shown by the paint marking—is destroyed. That eliminates the possibility of comparison by a single scout. If that kind of comparison is important, the scientists reason, then the swarm should be unable to decide where to go without it, or, at the very least, should take significantly longer to decide.

Although the experiments are still under way, Dr. Visscher said, there are hints that the findings will challenge existing notions about how swarms go about finding a new home.

Dr. Visscher expects the project to take several years and to open other avenues of inquiry. The idea that some scouts may dance for one site, visit another and then change their dance to favor the second site—something that in Dr. Visscher's earlier studies has been rare—has inspired some researchers to suggest that bees might be capable of forming a mental construct based on what they have seen.

At this point, Dr. Visscher is not sure that is possible. "Sometimes it seems as if bees can do anything," he said. "Do they really act as an automaton? Or is there some degree of awareness? It's difficult to think of an experiment to tell which an animal is doing. I wonder what it is to be a bee."

—DENISE GRADY, April 1997

Ant and Its Fungus
Are Ancient Cohabitants

THEY ARE TINY mandibled versions of Shiva, the Hindu god of devastation and restoration. In a mere three days, they can strip away every last trembling leaf, every vestige of chlorophyll from a large grove of trees. A herd of elephants or a blazing inferno could hardly do more damage to the face of a forest. Yet once they take their herbaceous plunder underground, the pillagers become gentle farmers, using the leafy matter to cultivate vast gardens of blooming fungi. They nourish the fungus, and the fungus in turn feeds their hungry multitudes.

And so the famed leaf-cutting ants act out their high drama of destruction and renewal, defoliating trees, bushes, vines, everything in their path—and from the wreckage creating a subterranean Eden, a myrmecian paradise.

The leaf-cutters represent the most advanced division of a powerful insect tribe called the attine ants, 200 species that engage in a mutually convenient arrangement with fungi. The ants and the fungi are symbionts, dependent on one another for survival and each having evolved specializations to optimize their intertwined existence.

The ants offer the fungi huge amounts of plant material that they gather from their surroundings, far more than the sedentary fungus could engulf on its own; and the fungus mulches the biomass and grows so fat that its hyphal tips swell with nutrients, sugars and protein, on which the ants can nibble.

Scientists have long been impressed by the harmony of the partnership between attine ants and their colluding mold. And what scientist could ignore the ants' spectacular gardens when in building them the insects displace enough earth to fill a good-size human living room?

Ant Farmers and Their Crops

Ants sever leaves from trees and return to nest within minutes, leaves are cut up into smaller and smaller sizes by smaller and smaller ants and are chewed up and implanted into living masses of fungus that line the nest chambers. Smallest members of the colony are in charge of caring for fungus plantation, weeding out foreign spores and periodically harvesting edible strands to feed the colony.

42

Michael Rothman

Yet only now are biologists discovering the nuances of the relationship and the evolutionary history behind it. They are applying molecular tools to reconstruct the genealogy of the symbiosis, determining when it arose and how it progressed over millions of years to assume, in its peak among the leaf-cutters, a partnership so powerful that it virtually controls the ecosystem of many regions of the neotropics.

Dr. Edward O. Wilson, a naturalist at Harvard University and author, with Bert Holldobler, of the acclaimed book *The Ants,* has described the adaptation of ants using fungi to take advantage of fresh vegetation as so successful "that it can be properly called one of the major breakthroughs in animal evolution."

In two papers appearing in the journal *Science,* researchers describe a number of complexities of the ant-fungal affair. They demonstrate that the coevolution of the attine ants and their fungi dates from 50 million years back, reaching varying degrees of codependency in each case. Among the leaf-cutters, the relationship turns out to be so tightly linked that the ants have relied on a clone of the same fungus for 25 million years, with each new colony sowing its first garden with a bit of starter fungus from the parent nest. This means that every fungal crop found on every leaf-cutter's farm throughout South and Central America, where the ants thrive, is a descendant of a single ancestral spore.

To the relief of mycologists, the researchers have solved a century-old mystery. Experts had not been able to determine the fungal variety through a traditional taxonomic analysis of the fungus's fruiting body—the cap of the mushroom, where the spores are found. In becoming a symbiont with ants, the fungus had ceased making fruiting bodies to reproduce itself and instead had come to rely on the insects to spread their seed around.

Now, by examining genetic sequences of fungal samples taken from the gardens of many of the attine species, the scientists have concluded that most of the molds are of the family Lepiotaceae, which claims among its members the parasol mushrooms familiar to connoisseurs and pickers and not all that different from the little white mushrooms sold in supermarkets everywhere.

The scientists also demonstrate that while members of the leaf-cutting branch of the attine tribe have become near-monoculturists, generally sticking with one very ancient strain of fungus to feed upon, those in another group of the ants are more supple in their agriculture techniques, occasionally acquiring new fungi from the outside to refresh their stocks and perhaps provide a variety of flavors in their diet. The more omnivorous ants are considered the more ancient or primitive types, closer to the founding fungal farmers that shopped around for the best spores to exploit in their gardens.

"This is what it must have been like fifty million years ago when this symbiosis was just getting started," said Dr. Ulrich G. Mueller of Cornell University. "It wasn't a single acquisition event, but rather something that happened over a period of time as the ants were driven into the role of associating with fungi."

The new work is of interest on multiple counts. Scientists now have a better understanding of the symbiosis between ants and fungi than they do of the other mutualistic arrangements between natural organisms, of which there are many. Mycologists celebrate the research for its emphasis on fungi, which are of fundamental importance to all ecosystems on land and yet which are so robustly ignored that most universities do not bother having a mycologist on their faculties.

"Fungi are more numerous than plants by sixfold, yet there are a tenth the number of people studying them," said Dr. Thomas Bruns, a mycologist at the University of California at Berkeley. "That's starting to change as ecologists recognize that fungi are the basis of all terrestrial ecosystems. These papers add a lot of wonderful new data to the fungal sequence banks." Mycologists emphasize that ants are far from the only organisms to discover the advantages of cooperating with fungi. Many plants, for example, rely on fungi growing at the base of their roots to take up needed nutrients from the soil.

Ecologists also point out that the research underscores the interdependency of life, serving as yet another reminder that habitats are composed of threads so tightly woven together that to yank out one stitch could lead to the unraveling of entire chains of life.

"Many organisms are involved in a community structure, and if an insult to the environment hurts or changes one element of that community structure, that could result in the loss of the capacity of the community to survive," said Dr. Mitchell L. Sogin of the Marine Biological Laboratory at Woods Hole, Massachusetts.

He also emphasized that microorganisms like fungi often are symbiotic with macroorganisms like animals, and that any discussion of biological diversity must include an appreciation for the invisible among us. Where would humans be without the microorganisms in their gut that help digest food, he said, or for that matter, without the mitochondria in their cells, the powerhouse structures in the interior of cells that supply energy for the body and that are thought to have once been free-living bacteria-like organisms?

Understanding the relationship between fungi and attine ants may also have practical applications. Leaf-cutters in tropical countries are essential to recycling plant material and to keeping the soil aerated, as earthworms do in more temperate climates. Yet they are also a menace to humans who try to farm in tropical regions of the New World.

"Leaf-cutting ants are the dominant herbivore in the neotropics, taking about 20 percent of all the fresh-leaf biomass there," said Ted R. Schultz, a graduate student at Cornell who was an author on both papers in *Science.* "It would be fair to say that the ants are the main reason it's hard to do agriculture in the neotropics."

Most of the plants native to the neotropics have evolved sufficient defenses that they are not entirely torn apart by the ineluctable energies of the attine ants, he said, but when people try to bring in foreign crops—fruit trees from California or Africa, say—the results are disastrous. "They'll strip entire orchards in no time," Mr. Schultz said.

North American farmers do not have the same concern about trying out foreign plant species, for although attine ants live as far north as Long Island, these species are of a less ambitious variety, and they feed their fungi less desirable goods, like insect feces or dead-leaf material.

In performing the current analysis, the researchers took genetic samples from 21 fungi isolated from the nests of 19 different types of attine ants and compared those genes with the DNA of free-living fungi not beholden to any ants. They also linked the genetic information from the fungi with previous analyses that had been performed on the molecular history of the attine ant tribe.

After subjecting the data to a series of elaborate statistical calculations, the scientists came up with phylogenetic trees showing divergences among the ant species and their fungal mates. The results led the scientists to conclude that the more primitive ants pick and sample new fungi from the environment, while the most highly specialized species remain unerringly faithful to the fungus that has served them so well.

The reason the leaf-cutters have stuck with their fungal strain is that the fungus can do what most fungi cannot: break down fresh leaves into usable nutrients. This means that a resource most herbivores find too daunting to nibble was laid open to the ants. Although incapable of digesting the leaves themselves, they could feed it to the fungus, allow the fungus to metabolize the material and then gorge themselves on the nutritious upwellings of the fungus. The beauty of the system has allowed leaf-cutters to evolve into the paradigm of what Dr. Wilson calls a "superorganism," a collection of individual ants that are each as specialized in their tasks as the cells of the body and that jointly perform the task of keeping the nest alive.

And what a nest it is. Those who study leaf-cutters in the tropics describe a leaf-cutter farm as a spectacle almost beyond belief. A mature nest may contain eight million ants, ranging in size from the tiny tenders of the fungal garden that are each no bigger than a letter on this page, to the pulsating, egg-swollen queen, which is the size of a whole, unshelled peanut. Most of the nest is underground, an elaborate warren of thousands of chambers ranging in size from a fist to a soccer ball, and all having been excavated by the ants. The chambers are filled with the spongy gray hyphae of the fungus that feeds all.

Aboveground, the area surrounding the nest is kept scrupulously clean, tended by janitorial workers that will whisk off any leaf, any interloper, any bit of undesirable mold that may slow the ceaseless passage of the cutter ants heading to and from the farm. The gatherers of plant matter, each about the size of a housefly, venture out to slice leaves into crescent-shaped bits, which they hoist back to the plantation.

In industrial, assembly-line fashion, the crescents are taken up by smaller ants that slice the slices smaller still, and then pass the pieces along to yet tinier ants that chew the material and soften it with enzymes into a moist paste. Even smaller ants then spread the paste like a slathering of jam over the fungal substrate. The mold then takes over and does the rest, threading its hyphae tentacles through the plant matter and breaking down the cellulose into nutrients the ants can use.

The fungus is pampered like a pet, fed to fatness and kept clean of competing microorganisms that might threaten the crop. As long as there are leaf-cutters, the lepiotaceous mold will thrive, for no virgin queen leaves home on the wing without first tucking a bit of it into her mouth to seed an estate of her own.

Atta cephalotes ants sever leaves from trees and return to nest. Within minutes, leaves are cut up into smaller and smaller sizes by smaller and smaller ants and are chewed up and implanted into living masses of fungus that line nest chambers. Smallest members of colony are in charge of caring for fungus plantation, weeding out foreign spores and periodically harvesting the edible hyphae to feed the colony.

—NATALIE ANGIER, December 1994

The Cicada's Progress

1. Cicada nymphs from the latest brood ascend a tree after 17 years in the ground feeding on the sap of tree rootlets. They have already gone through various molts, or instars, while in the ground and will shed again as they climb. They do not all grow at the same rate but are all ready to emerge at the same time. The trigger for their simultaneous emergence may be soil temperature, light or plant hormones, scientists speculate.

2. The adult emerges from the last nymphal case. It is still pale in color and the wings have not yet unfurled, filled out and stiffened.

3. The mature adult departs on a flight in search of a mate. The sexes can be distinguished by looking on the underside of an adult. The male has flaps for the famous cicada song. The female has an ovipositor, a spade-like tool for depositing the many eggs. After five and a half weeks, the hatchlings will fall to the ground like Ping-Pong balls and burrow into the soil for a 17-year stay.

Michael Rothman

Cicadas: They're Back!

IT WAS A MOST UNUSUAL coming-out party. The host wore a tan plastic rain-coat, not a dinner jacket. There was no bubbly music by the likes of the band leader Lester Lanin. And the swarms of gawky 17-year-olds? They were fashionably late.

For this was a gathering to celebrate cicadas, the inch-long insects that play out a three-week ritual of sex and death in the spring. The old generation crawls out from little tunnels in the ground and mates while flitting about. The new generation, the one Dr. Charles L. Remington had hoped to meet and greet as he tramped through the woods in the rain this morning, falls from hatching-places in the trees, bounces like tiny Ping-Pong balls on the forest floor and burrows in again.

From Connecticut to North Carolina unseasonably wet and cold weather seems to have delayed their once-in-a-lifetime entrance. Not to worry. Dr. Remington, a 74-year-old entomologist at Yale University, is a patient person. He has only been waiting for this party since 1979.

Yes, it has been 17 years since this batch of cicadas made their last appearance, 17 years since their parents droned all day in a conversation-stopping chorus—a synchronized love song.

It has been 17 years since people complained that they sounded like a jet engine revving up for takeoff, 17 years since they crunched underfoot after dying, 17 years since they sent panicky gardeners diving for the encyclopedia.

"Arguably this is the most remarkable insect in the world," Dr. Remington declared.

Dr. Remington is in a minority: He loves cicadas. Farmers in the seventeenth and eighteenth centuries did not. "They thought a locust plague had been visited upon them," said Dr. Louis M. Vasvary of the Rutgers Cooperative Extension in New Brunswick, New Jersey. And in 1970, cicadas so unnerved Bob Dylan that he wrote a song called "Day of the Locust." The inspiration was a sleepless June night of trilling and wing-flapping in Princeton, New Jersey.

He was not troubled by the tuh-MAY-toe/tuh-MAH-toe problem that bedeviled the Gershwins: Cicada is pronounced with a long a, sih-KAY-duh. But Mr. Dylan confused cicadas with locusts, which are chewing insects that can wreak havoc on crops.

Cicadas are suckers. They spend most of their lives as a kind of underground parasite, drawing in fluids from tree roots. True, female cicadas jab at tree leaves when laying their eggs, but the trees sustain no permanent damage. And the new generation of cicadas soon falls to the ground and burrows in, out of sight and out of mind for 17 years.

Cicadas have been documented since the days of the Pilgrims. But Dr. Remington points out that scientists still know relatively little about why cicadas are the way they are. Among the unknowns: Exactly why do they emerge every 17 years? Why do they live longer than virtually any other insect? How many cicadas inhabit a single colony? How many eggs can a female cicada lay? And are starlings threatening them with extinction?

Dr. Remington maintains that their life cycle is regulated by what are essentially two body clocks. One, which he believes is on the cicada's subesophageal ganglion, ticks off years. When it hits 17, the second clock takes over. It tells the cicadas to withdraw their beaks from the tree roots and start crawling upward.

Dr. Remington speculates that the second clock is a "photomeasurement device" that can sense light. But he has no easy explanation for how it is activated while the cicadas are still underground. Nor is it clear how it could be timed precisely enough to send thousands of cicadas pouring out of their tunnels by the millions almost simultaneously.

Dr. Chris Simon, an evolutionary biologist at the University of Connecticut at Storrs, suggests that seasonal changes in plant hormones or amino acids in the cicadas' tree-root diet may affect the countdown. In a 1995 paper in *The Annual Review of Entomology,* she wrote that some entomologists believed that the trigger was soil temperature—when it warms up and the cicadas' bodies reach a certain temperature, they are programmed to emerge.

In fact, some gardeners think the ground-temperature threshold is 64 degrees. But Dr. Remington discounts that idea. "If the soil temperature is the critical factor that lets them out, that's terrible for cicadas," he said. "Even in one little valley, there are cool places and warm places. That could bring them out raggedly instead of in overwhelmingly large numbers, and the predators would wipe them out."

Dr. Simon has found that the nymphs, as the cicadas are called when they are underground, do not all grow at the same rate. Yet they are all ready to emerge at almost the same moment when the 17 years end. Some have been waiting in exit tunnels for as long as a year.

Dr. Carolyn Klass, an extension entomologist at Cornell University, said they apparently traveled up and down in the tunnels. On 95-degree days last month, gardeners in Rockland County found them almost at the surface. But on cooler days they retreated to about four inches below the surface and waited another week before emerging.

What is not in dispute is that cicadas have the longest juvenile period and one of the longest life cycles of any insect discovered. Some termites live 10 to 12 years, but they whiz through adolescence and settle into a long adulthood. Hellgrammites, aquatic insects with sickle-shaped mandibles found in the eastern United States, spend three years in an underwater phase that Dr. Remington said is roughly comparable to the cicada's near-hibernation. At the opposite end of the spectrum, the shortest known life cycle is in the fruit fly, which lives only 21 days.

The cicadas in the eastern United States are on a 17-year schedule; cicadas in the South and Southwest are on a 13-year schedule. Dr. Remington said they are essentially the same insect. Is there any significance in the fact that 13 and 17 are prime numbers? "We have no reason to think it means anything," Dr. Remington said.

Biologists long ago realized that, each year, a different batch of 17-year cicadas emerges, and they set up a system with 17 Roman numerals to keep track of the different broods.

As it happened, no cicadas emerged on the 14th, 15th and 16th years, Dr. Remington said. "It turned out there absolutely aren't any, and two of the documented broods have gone extinct." There are now a dozen broods and each brood varies genetically and geographically from the others, he said. This year's cicadas are Brood II.

But how many species are there? Dr. Remington argues that there should be six, three for the 17-year cicadas and three for their shorter-lived look-alikes. Dr. Simon argues for three, with no differentiation between the two varieties.

Dr. Remington is watching for the mass exodus, the day birds can feast on cicadas until they are stuffed—and are still outnumbered by insects. But Dr. Remington is worried that starlings, which he said were not imported to the United States until 1915, may spoil the party. New York City passed a law banning pet shops from selling them a few years after they arrived, and to get rid of birds that were suddenly unmarketable, "people let a bunch of them out," Dr. Remington said.

He suggests that starlings may be implicated in the disappearance of 35 of the 75 cicada colonies that were known in Connecticut around the turn of the century.

"That's a scary decline," he said.

Dr. Klass of Cornell went a step farther. "My sense is, each 17 years there is probably a decrease in the locations and the numbers because of people pressure," she said. "We're moving into land that was once fields and before that, wooded areas. As we push out, we are paving over and changing the habitat, so that if you look far enough in the future, probably they're going to die out."

Clearly, many cicada colonies have been lost to development. Development in Nassau County since 1979 will probably cut into cicada numbers on Long Island this year, said Daniel Gilrein, an entomologist at the Cornell Cooperative Extension in Riverhead, Long Island. Nymphs that burrowed into the ground 17 years ago would have been entombed if the ground was paved over, he said.

That is not a problem in Dr. Remington's cicada preserve in—appropriately enough—Sleeping Giant State Park, about nine miles north of the Yale campus. But like Linus in the pumpkin patch, he is still waiting for the throng he was counting on counting. One of his goals was to determine exactly how many eggs female cicadas lay.

"I'm looking forward to two more colony hatches," the 74-year-old scientist said, counting ahead to the year 2030. "I became an emeritus professor a few years ago, and since then, I've been teaching more than I ever did. I've got too much to do in the interim."

—JAMES BARRON, June 1996

"Cicadas are among the world's most palatable insects," according to Dr. Remington. He was talking about how 17-year insects with greasy-looking wings taste to predators like birds. He plans a taste test later this year, for blue jays and starlings.

But he is also collecting cicadas from the Sleeping Giant preserve and plans to boil one or two so that he can taste essentially what the birds taste. He said that in a week or two, he would serve guests stir-fried cicadas.

Dr. Remington noted that insects are on the menu in northern Japan, in inland provinces where people do not get enough fish and meat protein. The favorite food insect there, he discovered in World War II, is the yellow jacket. But he said that adult cicadas were also served.

Tasty 17-Year Cicadas

Serves 2 amply

1 cup freshly emerged cicadas
2 quarts clean boiling water
salt to taste

1. Gather cicadas from tree trunks and shrubbery, just after they have come out of their nymphal shells; they should still be soft and whitish, like softshell crabs.
2. Drop the cicadas into water after it has come to a full boil; water may be salted.
3. After 12 minutes, drain and season to taste.
4. As a variation, try older cicadas, 30 to 60 minutes after they emerge; they are still tasty, but have hardened and darkened. They should have their wings and legs snipped off after boiling.
5. For still another variation, gather nymphs while they are still living underground or just after they emerge.
6. Boil as above.
7. Use boiled nymphs in much the same way you would use cooked shrimp. For example, stir-fry in a wok, combining with favorite spices, vegetables and sauces.

Nutritional analysis: High in protein, fat and glycogen, the form in which sugar is stored to provide energy.

Dance of Love and Death
Mating scorpions engage in a long and violent waltz, locked together at the mouth and front legs. The male repeatedly stings the female and the female thrashes about. Sometimes the heavier female consumes her partner after mating. Chemical signals control each step of the elaborate ritual.

Achieving Fertilization
A closeup of the locked position of mouthparts and pincers during a kind of massage that allows the male to position the female for sperm transfer.
Researchers speculate that the massage may help suppress the aggressive tendencies of the female.

54

Bob Ziering

The Scorpion, Bizarre and Nasty, Recruits New Admirers

TO THE ANCIENT CHINESE, snakes embodied both good and evil, but scorpions symbolized pure wickedness. To the Persians, scorpions were the devil's minions, sent to destroy all life by attacking the testicles of the sacred bull whose blood should have fertilized the universe.

In the Old Testament, the Hebrew King Rehoboam threatened to chastise his people, not with ordinary whips, but with scorpions—dread scourges that sting like a scorpion's tail. The Greeks blamed a scorpion for killing Orion, a lusty giant and celebrated hunter.

Throughout history and across almost every cultural boundary, scorpions have had a rotten reputation. And as far as Gary A. Polis, one of the leading authorities on the creatures, is concerned, they deserve it.

"There's a good reason why humans are terrified of scorpions," he said. "They've caused a lot of deaths over the years."

And even those species whose venom is relatively innocuous, he said, can deliver stings that "feel like 10 flaming bullets rotating inside you all at once." As though the description needed reinforcement, he rammed his finger against his chest and gave it a twist.

But Dr. Polis, an associate professor of biology at Vanderbilt University in Nashville, is not a biblical doomsayer redux. He is one of an elite and growing cadre of researchers here and abroad who are dedicating their careers—and braving nature's version of Uzi fire—to the study of the strange lives, violent nights and brutal loves of scorpions, the nocturnal relatives of spiders and other members of the eight-legged invertebrate family known as the arachnids.

Long neglected in favor of spiders or their distant six-legged cousins, the insects, scorpions are finally winning researchers' attention and respect.

Devotees say that, while scorpions may be among the most ancient of terrestrial animals, they have features that make them seem like the most modern of mammals. And only lately have scientists realized how singular
scorpions are.

"We know a lot about insects and crustaceans, but we're just beginning
to find out about arachnids in general, and scorpions in particular," said Dr.
Philip Brownell, a biologist at Oregon State University in Corvallis. "Scorpions are an extremely successful group of animals. They use all sorts of
bizarre sensory systems to find their way around the environment."

Many of the details of the scorpion's peculiar traits and pastimes have
been gathered into a book, *The Biology of Scorpions,* edited by Dr. Polis and
published by Stanford University. It is the first attempt to synthesize all that
is known about the creatures, and the portrait that emerges reads like the
invertebrate edition of Guinness: scorpions are some of the biggest, mean-
est, longest-lived, most sensitive, most maternal, least fraternal, slowest,
quickest and certainly the most weirdly colored creatures among the arachnids and insects.

"When you talk about scorpions, you tend to use a lot of words like
'the only known example,' 'the first,' 'the largest,'" said Dr. Polis. "It's just
one gee-whiz fact after another." Many of the new findings startle even old
scorpion hands. Biologists from Frankfurt, for example, have discovered
that one of the largest species of scorpion, which is found on the Ivory Coast
of Africa, is social to a degree unheard-of among arachnids, which normally
are solitary creatures. Males and females, which can weigh almost three
ounces apiece and measure up to eight inches in length, live together and
rear their young for two years or longer.

In caring for their offspring, the adults will kill rodents, frogs and other
vertebrates, strip the prey apart, grind it up and feed the predigested stew
to their young.

But scorpions are not always model spouses and parents. Researchers
who follow the creatures out in the field have learned that some species are
among the most aggressively cannibalistic of all creatures, deriving 25 percent of their energy by consuming their neighbors, their mates, even their
own young.

In areas where more than one species of scorpion compete for
resources, the creatures engage in elaborate interspecies feasting that would

make the Borgias look like the Cleavers, with the elder members of the smaller species eating the offspring of the bigger species, the bigger species in turn devouring the more diminutive adult scorpions, and two adults of similar sizes clashing for the right to remain, however fleetingly, at the top of the food chain.

Researchers have gained new insights into scorpion mating, among the nastiest affairs in nature. Males and females engage in lengthy and violent waltzes, moving to and fro, to and fro, front legs gripping front legs, mouthparts locked together and tails whipping forward, as the male repeatedly stings the female and the female thrashes about, seemingly furious at being dragged around.

Sometimes, after copulation is completed, the female, which is almost always heavier than her mate, will exact revenge for the ordeal by consuming her partner.

Scientists are deciphering the chemical signals that control the scorpions' behavior at each step of the intricate mating ritual. They have also gathered evidence that the reason for the extended dance is to allow the female to assess the male's genetic worth before accepting his sperm.

Neurobiologists are beginning to appreciate the scorpion's brain, which offers them a rare combination of simplicity and complexity. Unlike other arachnids and insects, which have their nerve cells disseminated up and down their body, a scorpion has a cluster of neurons in its head, just as a mammal does. But a scorpion's brain has far fewer nerve cells than a mammal's, making it easier to study.

It is not only scientists who are interested in scorpions. As a result of the Persian Gulf crisis and the stationing of American soldiers in Saudi Arabia, the Pentagon is, too. The Defense Department recently asked Dr. Polis and other scorpion experts to travel to the Middle East to study several of the potentially deadly species found there.

Dr. Polis is considering accepting the Pentagon's offer, but he and his colleagues doubt that the scorpion threat to American troops is so pressing. They say excellent antivenom medications are available to prevent death or even serious illness. Pain, however, is another matter.

"For soldiers in the Mideast, it's rather important to make sure every morning that no scorpions have crawled into their shorts or pants," said Dr. Dean D. Watt, a biochemist at the Creighton University School of Medicine in Omaha, Nebraska. "The scrotal area is highly vascularized, so if you happen to get stung in that area, it may not kill you, but it will hurt a lot worse than it would on the finger." The pelvic network of blood vessels, he explained, permits the excruciating pain from the venom to spread across the legs and torso.

Lest anybody think them odd for spending their days and nights on one of humanity's ancient foes, scorpion mavens emphasize the many pleasures of their chosen specialty. Not only are scorpions intrinsically bewitching, they say, but the best places to study them also happen to be the most exotic locations. "They live in absolutely beautiful environments," Dr. Brownell said. "I have the built-in excuse to travel around the world."

But scorpion specialists will not always own up to their obsession. "I don't like to talk about my work at the average cocktail party in Nashville," Dr. Polis said. "I'll go through six layers of explanations before I mention scorpions. People will probe and probe, and when I finally tell them what I study, they say, 'You do what? What kind of weirdo are you?'"

Dr. Polis and others insist they pursue their studies for the most rational of reasons—because scorpions are an ideal "model system" to approach the scientific questions that interest them. "If you ask me do I love scorpions, I'd answer a flat no," Dr. Polis said. "I love the information I get from them."

Naturalists from Aristotle on have been mesmerized by scorpions, but only within the last two decades have researchers had the wherewithal to extensively study the animals in the field. The invention that has revolutionized scorpion biology is the portable ultraviolet light. Another of the scorpion's exceptional features is its ability to glow under ultraviolet light like a psychedelic poster.

The exoskeleton of the scorpion is made of a tough layer of tissue that feels like fingernail but is in fact chitin, a type of cuticle protein. The coat reflects back the ultraviolet rays from moonlight and other light sources so brightly that even a black scorpion glows a fluorescent shade of green or pink. Fossilized scorpions from 300 million years ago still gleam brilliantly under ultraviolet light.

Scientists are not sure why scorpions fluoresce, although some suggest that the glow evolved to attract insects, which are drawn to ultraviolet light. Whatever the reason, the unmistakable shine, visible from 20 feet away, makes it easy to spot scorpions in the dark, when they emerge to eat, mate, fight and otherwise carry on.

"With UV light, we can locate them, capture them, mark them, let them go and recapture them," said Dr. Neil Hadley, a zoologist at Arizona State University in Tempe. "We can study their metabolic rate, their oxygen consumption, anything we please. UV light has really changed the way we study scorpions."

Through the use of ultraviolet lamps and other technologies, researchers have learned that scorpions have changed little since the Silurian epoch, 400 million years ago, when they were among the first animals to crawl from sea to land.

Scorpions are commonly associated with the desert but in fact the 1,500 known species are found in almost every possible setting: deserts, rain forests, savannas, grasslands, temperate forests. Blind ones creep around caves half a mile underground; tiny ones burrow in the cracks of pineapples; stout ones cling to the slopes of the Himalayas 14,000 feet above sea level. Scientists believe that 500 to 1,000 other species remain to be discovered. "If you go someplace where few other scientists have been, there's a good chance you'll find a new scorpion," said Dr. W. David Sissom, a biologist at Elon College in North Carolina.

All known species are predatory and are equipped with venomous stingers, although only 25 species pack enough toxin to kill a human being. The venom is carried in a gland on the back of the tail, and the animal can whip its stinger forward in a fraction of a second to jab a victim, sometimes repeatedly. Chemists have determined that the venom is a brew of up to 30 neurotoxins, each designed to fell a different type of prey. Some of the neurotoxins have been found to be most effective against insects, while others are best at paralyzing frogs and other small vertebrates.

Once a prey has been knocked out, the scorpion begins the lengthy business of liquefying its victim. Like spiders, scorpions digest their food before consuming it, spitting out enzymes to dissolve the prey into a broth that the scorpion can suck into its mouth. Scorpions have more in common with spiders than a digestive style. When they live in the same neighborhood, the two arachnids compete for the same resources, the insects.

But scorpions have a distinct advantage over their competitors: that is, a taste for them. And given their usually superior size, they can usually turn their competitors into prey with little fear of reprisal. In areas where scorpions abound, spider populations are generally kept in check.

But a scorpion is by no means immune to predation. Although it can thwart some potential attackers with its venom, it is so meaty that owls, bats, snakes and other animals will endure the sting for the sake of a hearty meal.

Assuming they avoid being consumed, scorpions have the potential to live 15 to 25 years and perhaps beyond, longer than any other known arachnid or insect. Contributing to that longevity is the scorpion's miserly metabolic rate, which is slower than that of any other invertebrate. Creatures with slow metabolisms generally live far longer than those that burn energy at a rapid clip, as most small animals do.

"It's been calculated that a scorpion has a metabolism equivalent to a growing radish root," Dr. Polis said.

Scientists have learned that scorpions possess such a sluggish metabolism to allow them to survive in extremely harsh conditions of heat and cold on virtually no food or water. They can live for more than a year without eating, and they are covered with a slick of wax that seals in water. Even in urinating or defecating, they conserve water, releasing nothing but a powder of waste products.

Everything about the scorpion turns out to be extended in time. They take up to seven years to mature, and they gestate their young for up to a year and a half, a pregnancy rivaled only by the elephant. More surprising still, scorpion mothers have something like a mammalian placenta, which nourishes their young internally, another feature unique among invertebrates. The offspring are born live, and then crawl onto their mother's back for another two to six weeks of external development.

Those in the scorpion business say they are most impressed by the animal's exquisite sensitivity. Everything about it is designed for detecting and capturing prey in the darkest night. New studies suggest that the animals navigate by starlight.

Other scientists have discovered that scorpions are ambulatory seismographs. On their eight legs are slit-like organs that can sense surface disturbances from an insect walking on sand as much as three feet away. "That's really surprising because sand is supposed to damp compressional waves," Dr. Brownell said. "But the scorpion can sense the microearthquake of a walking insect, and it will run toward it."

The pincers of many species are covered with ultrasensitive hairs that, by vibrating at different speeds and in different directions, tell a scorpion that a flying insect is approaching, allowing the predator to snatch the prey from the air.

As sensitive as they must be to eat, scorpions also have senses tuned to mating, and to escaping. In recent studies of male scorpions, biologists have determined that two strips of sense organs running down the middle of the animal's chest, called pectens, can sense seductive pheromones left behind by just a single foot of a female.

The male's pectens also seem to help him find a vital stage prop during the mating dance: a stick upon which he can deposit his sperm packet. During the dance, the male must drag the female over to the stick, release his sperm, and help position her over the stick. Eventually, the female will open her genital slit, located between her own pectens, and aspirate up the sperm packet.

The male's pectens may also help tell him, through chemical signals, that a female has completed intercourse, and is on the verge of attacking him for her breakfast. He will immediately try to break away from her and escape, but about 10 to 20 percent of the time he fails and is eaten.

Indeed, most scorpions are so notoriously hostile toward one another that many scorpion specialists are eager to learn more about the handful of giant species of scorpion that live in relative harmony with one another. "These are scorpions that live in colonies, which don't show the usual degree of aggression, and there doesn't seem to be any cannibalism among them," Dr. Sissom said. "These are exceptions to the usual rule of scorpions."

Dr. Polis suggests that the cooperative species might have some kind of colonizing pheromone that tells the scorpions not to attack each other and to be more social. Some researchers are now looking for that tranquilizing chemical.

The notion of a cooperation pheromone is sweet, but even if scientists manage to isolate the substance, it can only be hoped that the scorpion retains its reputation as a ruthless night fighter, irrepressible cannibal, sexual athlete and devil's handmaid. If nothing else, the scorpion, as the Methuselah of invertebrates, just may outlast any efforts to tidy up its delicious bad name.

—NATALIE ANGIER, NOVEMBER 1990

Defenders Rally to Misunderstood Robber Fly

ROBBER FLIES are neither pests nor pollinators. They bite neither man nor beast, and they do not spread disease. Yet despite this detachment from any realm of human affairs, they have become the central passion of Dr. Robert Lavigne, a biologist at the University of Wyoming in Laramie.

Some 30 years ago, while helping to survey the parasites and predators of Wyoming's grasshoppers, Dr. Lavigne (pronounced la-VEEN) became so intrigued with the predatory stealth, aggressiveness and occasional cannibalism of robber flies that he has remained engrossed in their study ever since.

His research is not exactly a stimulating subject for cocktail party conversation. "Most people think entomologists are strange no matter what they do," he remarked. Nor is his research likely to net him a Nobel Prize. But he has already attained the luster of having had a new species named for him, *Robertomyia lavignei*. The one-and-a-half-inch long grayish speckled insect is one of three new species he discovered in a recent three-year stint as an agricultural aide in Somalia.

Worldwide, nearly 5,000 species of robber flies are known and thousands more await discovery, Dr. Lavigne said in an interview. "Why, every year, 100 to 200 new species of robber flies are being described. They are a prime example of the tremendous biodiversity around us."

Wyoming alone is home to more than 100 species of robber flies that live everywhere from high plains deserts to mountainous coniferous forests. Their behaviors are also rather complex for an insect and thus repay the decades of study by Dr. Lavigne, who specializes in ethology, the science of animal behavior.

Some robber flies are mimics that, like the proverbial wolf in sheep's clothing, can sneak into a colony of their favorite prey and pick off unsuspecting neighbors one by one.

The species that mimics bees, in fact, gave robber flies their common name. This black and yellow bee-like fly dwells unmolested on and around honeybee farms and gobbles up worker bees as they return laden with pollen to their hives. Beekeepers in Germany some 200 years ago understandably dubbed the flies thieves.

The males of many robber fly species court their mates, although others that rank lower on the evolutionary totem pole do not. By devoting his career to the study of this one insect family, Dr. Lavigne said he has been able to shed light on the evolution of courtship, a rather sophisticated form of insect behavior. In general, courtship is rare in the insect world and the few male insects that do solicit mates—lightning bugs by flashing, crickets and grasshoppers by calling—are well-known exceptions.

The eating habits of robber flies range from a limited menu of mostly two species to a cosmopolitan smorgasbord, including some insects that are troublesome agricultural pests.

A robber fly with an eclectic appetite might, for example, consume beetles, flies, bees, wasps, grasshoppers, dragonflies, stone flies, mayflies and thrips, among other winged creatures. A few dine on spiders, some of which may in turn dine on robber flies. And the females of some robber fly species are cannibalistic, albeit unintentionally so, and will devour potential mates that come too close to the larger female's grasping toes and piercing mouthparts.

"Courtship behavior seems to have evolved as a way of reducing the male's risk of being eaten by a female of his own species," Dr. Lavigne has concluded. He explained that cannibalism is almost exclusively a female trait that apparently results from a case of mistaken identity, since a hungry robber fly considers any insect smaller than itself to be potential source of sustenance.

"The female's attack at prey is so sudden that the cannibalism is really indiscriminate," he said. But by courting the female before getting close enough to mate, a male with amorous intentions in effect announces that he is neither predator nor prey but merely interested in propagating the species.

The male of one Wyoming species, for example, will hover near a receptive female and kick his legs, flashing patches of white and black hairs to dazzle his intended. In some species, the female responds in kind, for example, by spreading her wings and vibrating her legs to indicate her willingness to copulate.

But more often, female robber flies play hard to get. While the male ferociously vibrates his wings and elevates his body in a precopulatory gesture, the female may fly away or may vibrate her own wings and rise up on her legs to discourage a union.

Robber fly adults live only about four to six weeks. Most species overwinter underground or in logs as larvae. Like the adults, the larvae are carnivorous, often dining on grubs and other underground pests. Females lay their eggs in a wide range of habitats, from leaves to logs, and when the larvae hatch they burrow into protected environments with wholesome food supplies.

The honeybee mimic, for example, crawls along logs until she locates the tunnels of carpenter bees. She deposits one egg in each hole and when the larvae hatch they crawl down into the tunnels and, it is assumed, dine on carpenter bee larvae.

Dr. Lavigne and his associates at the university have stalked robber flies throughout Wyoming and on two continents abroad. His studies require the kind of patience foreign to most frenetic urbanites. For example, individual flies may be marked with pinhead-sized dots of model-airplane paint, in effect giving the animals name tags. Then the scientist may sit motionless for hours to observe the behavior of a single fly without disturbing it. While the fly forages, feeds, rests, mates and lays eggs, the scientist merely sits and watches.

When Dr. Lavigne extended his studies to the robber flies of Australia, his wife, Judith, served as an unpaid research assistant. On one occasion, she sat so still watching robber flies on a sand dune that a resident blue-tongued skink—a reptile about a foot and a half long—slithered up to make her acquaintance.

But as a rule, outside sources of excitement are uncommon in robber fly research. In 30 years of fieldwork, Dr. Lavigne said he encountered no threats to life or limb. Rather, the thrills of a field biologist lie primarily in unraveling the habits of a life form and determining its place in the scheme of nature.

Some robber flies capture prey while in flight, and others may spring as far as 20 inches from a perch on a plant or log. Most then bring their captives to a feeding station, the insect's version of a dinner table.

When a robber fly feeds, Dr. Lavigne explained, it first impales its prey on its fierce proboscis, a straw-like mouthpart, and injects a toxin that paralyzes the prey in 30 to 50 seconds. Next, it injects enzymes that digest the prey's muscle and fat tissue, then sucks up the juices.

Sated, the fly pushes the hollow husk of the prey off its feeding table. The undelectable morsel is then carried off and further consumed by ants, who leave behind only the insect's external skeleton to enrich the soil.

Dr. Lavigne has even explored the sounds made by robber flies during their various activities. With a microphone mounted on the end of a long pole, he and his research associates have bugged the flies on their daily rounds, and established that the insects have a repertory of different buzzes for each activity.

Some scores of biologists have now enrolled in the select fraternity of robber fly aficionados. But Dr. Lavigne remains the world's leading repository of data on these insects, having collected, in a dozen or more languages, nearly all research reports ever published worldwide on the insect.

—JANE E. BRODY, August 1992

Mosquitoes' Tricks
Still Exceed Remedies

THE PERPETUAL WHINE around one's ears and the repeated slaps at the persistent airborne pests were like a retrospective weather report. It was precisely 10 days after a major thunderstorm had drenched the St. Croix River valley and the latest opportunistic crop of those infamous summer spoilers—mosquitoes—had hurried through their transformation from egg to larva to pupa, producing bloodthirsty adult females intent on guaranteeing a next generation.

Here in Minnesota, where T-shirts sporting chest-sized mosquitoes call the insect the "state bird," it is little comfort to know that to a mosquito, human blood is second-rate, less tasty than that from birds and even buffalo.

But at least one Minnesotan, Dr. Roger D. Moon, an entomologist, says the time has come to appreciate this adaptable creature for what it is. The world's 3,500 species of mosquitoes—51 of them in Minnesota and 57 in New York State—live in deserts, rain-filled hoofprints, tree holes, pitcher plants, puddles created by melting snow, old tires, empty cans and flower pots as well as in more traditional wetlands. Any animal with blood—even those with thick fur or feathers, tough hides or limited skin area, like turtles—is likely to be attacked by one or more mosquito species.

"Mosquitoes have survived because they have evolved different ways to make a living, to get what they need to reproduce," explained Dr. Moon, who has been studying the subject for more than a decade. Understanding the evolution and biology of mosquitoes and the ecology of human-mosquito interactions is crucial to preventing the social and environmental devastation that can result from a surge in mosquito-borne diseases.

The malaria-bearing anopheles mosquito still resides in the United States and has caused recent outbreaks in California. The tree-hole mosquito, *Aedes triseriatus,* is a vector of La Crosse encephalitis, a brain infection with a mortality rate of about 30 percent. The mosquito picks up the virus from infected birds. Although the virus replicates in birds and mosquitoes, it does not make those hosts sick. But people bitten by a virus-carrying mosquito can become fatally ill, even though humans are dead-end hosts in which the virus does not reproduce well enough to help it survive, Dr. Moon said.

Even more worrisome is the recently introduced Asian tiger mosquito—named for its stripes, not its ferocity—which is proving to be an indiscriminate incubator for human pathogens. In the South, it has transmitted dengue, also called breakbone fever because it makes people feel as if their bones are breaking, and perhaps Eastern equine encephalitis, which has a fatality rate of 50 to 80 percent and leaves half its survivors with brain damage.

The Asian tiger mosquito, *Aedes albopictus,* although primarily a tropical species, has already adapted to Chicago winters by evolving changes in its biological clock to help protect it from cold. Dr. William A. Hawley, working at the United States Centers for Disease Control and Prevention in Atlanta, showed that the eggs of temperate-zone tiger mosquitoes, instead of hatching after they are laid, go into diapause—the insect version of hibernation—as winter approaches. That happens earlier in the fall in more northern habitats, protecting the eggs from early frost, because the insect's biological clock triggers diapause at longer day lengths as the species spreads northward.

"In only a few years, the mosquito's biological clock had evolved to track local climatic conditions," Dr. Hawley wrote in *Natural History* magazine. The species even briefly took up residence in Minneapolis, where winter temperatures of 20 degrees below zero are not uncommon. But the breeding population was wiped out by quick local action: a blast of insecticide and the elimination of a huge lot of old tires imported from Japan for retreading.

"We should have done the same all over the country when the Asian tiger mosquito was first detected," said Dr. George Craig, an entomologist at the University of Notre Dame and a specialist in mosquito-borne diseases. "But the threat wasn't taken seriously and now this mosquito, which acts like a natural test tube for deadly viruses, has become a common pest, infesting every county in Georgia and Florida, many Southern and Midwestern states and now in the East, in Delaware, Baltimore and Pennsylvania."

In addition to its talents for breeding viruses, the Asian tiger mosquito is an eclectic daytime feeder, capable of creating repositories of life-threatening infection in birds, rodents, farm animals and pets, increasing the chances of spreading disease to people.

Fortunately, no mosquito can transmit HIV, the virus that causes AIDS. Instead of multiplying in mosquitoes, as encephalitis viruses do, HIV is digested in the mosquito's gut, so there is no virus available to contaminate the saliva mosquitoes inject when they bite.

Dr. Moon put it this way: "If you are sitting next to an infected person with a swarm of mosquitoes all around you, you'd be likely to die of blood loss before a mosquito would spread the virus to you."

The nation's most common mosquito pest, appropriately named *Aedes vexans,* is just a nuisance and not a disease vector, Dr. Moon noted while he and his visitor strolled through a wooded area on the University of Minnesota's St. Paul campus. It was not long before the subject of his discourse alighted on his arm. He allowed the mosquito to bounce around like a lunar module looking for a suitable landing site until she found a hairless spot on his hand. Dr. Moon said, "When a person slaps a mosquito and says triumphantly, 'Got him,' it's really, 'Got her.' The males don't even bother to land on you."

The mosquito's proboscis quickly drilled into a surface capillary and she began filling her belly with blood. As she fed, she salivated, injecting an anesthetic and anticoagulant into Dr. Moon's hand to help insure an uninterrupted meal. A long minute or two later, the satiated mosquito, her abdomen red and swollen, drifted off, almost too heavy to fly, to a sanctuary where she could concentrate the protein in the blood and disgorge the watery remains.

No one knows why mosquitoes may appear to prefer one person over another. Carbon dioxide, the gas prominent in exhaled air, acts as a distant attractant, Dr. Moon explained, and mosquitoes have sensors that detect color and motion. They prefer dark colors and are sensitive to odors, Dr. Moon said, adding, "If the right olfactory cues are not there, the mosquito will get up and search elsewhere." Laboratory experiments have shown that mosquitoes prefer warm body parts to cold ones, that they are drawn to lactic acid in sweat and breath and that they respond to certain hormones and amino acids in body fluids. Menstruating women are avoided.

Experts say that there is little evidence that mosquitoes are drawn more to some people than to others. Some people react more vigorously to the allergenic ingredients in mosquito saliva. In fact, in most people, mosquito bites resemble desensitizing allergy shots; as the mosquito season progresses, victims react less and less to the bites.

Different species use different strategies to find their hosts. According to studies by a Japanese zoologist, Dr. Masahiko Nishimura, some are patrollers, who fly around looking for blood-filled targets, while others are pouncers, who wait for unsuspecting passersby. "These are the same two strategies large predators use to find their prey: stalking and ambush," Dr. Moon pointed out.

Female mosquitoes mate only once, storing a lifetime of sperm in an abdominal sac called a spermatheca. In most species, the females must then take in a blood meal to produce 125 or so eggs, which are fertilized en route to being laid. The process can be repeated one or more times over her lifetime. The eggs can survive for years in dry areas, waiting to hatch into larvae when the site becomes flooded. The larvae feed on microorganisms on the bottom of their watery home.

In most species, the larvae must surface periodically to breathe by punching through the water's surface with a respiratory siphon at their rear end. Once near the surface, mosquito larvae are easy pickings for such wetland wildlife as minnows and small fish, frogs, reptiles, wading birds, pregnant ducks, dragonflies and an aggressive carnivorous insect called a backswimmer.

The cattail mosquito's larvae stay in relative safety near the bottom. They use an armored siphon to punch holes in cattail roots and breathe air bubbles through them. This common mosquito, an early evening feeder, is thought to be the East Coast vector for Eastern equine encephalitis, Dr. Moon said.

Adult mosquitoes are eaten by bats and birds, so they are important in the food chain. And since all male adult mosquitoes and newly emerging females rely on plant nectar for sustenance, they are important pollinators, particularly for certain orchids.

According to Dr. Erik Kiviat, executive director of Hudsonia, an environmental organization at the Bard College Field Station in Annandale-on-Hudson, New York, when communities begin mosquito-control programs, they should consider the value of these insects as pollinators as well as the impact of control measures on beneficial species. "We need to reduce nuisance mosquito bites and disease transmission with a minimum impact on wetlands, birds, amphibians, fishes, pollinators, plants and other organisms," he wrote in the organization's newsletter.

Dr. Kiviat said two popular methods of mosquito control were nontoxic or relatively so to other animals and plants: methoprene, a growth-regulating hormone that disrupts the development of mosquito larvae, and *Bacillus thuringiensis israelensis,* or Bti, a bacterium that kills larvae when they consume it along with food particles.

Another biological technique involves introducing fish that preferentially dine on mosquito larvae, although this method has sometimes backfired when the fish drove out other desirable species, like frogs and salamanders. Dr. Kiviat said more promising biological controls might use smaller predators and parasites of mosquito larvae, like fungi, protozoa, nematodes, tadpole shrimp, dragonflies and—for the supreme insult—the cannabalistic mosquito, *Toxorhynchites.*

—JANE E. BRODY, August 1994

A Wealth of Forest Species
Is Found Underfoot

ENVIRONMENTALISTS have focused great attention on spotted owls and ancient trees in the forests of the Pacific Northwest. But a small cadre of researchers who have been studying insects and other invertebrates of the forest soil now say that the old-growth forest has been hiding underfoot perhaps its most astonishing biological secret.

Detailed studies of arthropods, including insects, spiders, mites and centipedes, in the soil of the old-growth forest suggest that the soil under the region's forest floor is the site of some of the most explosive biological diversity found on Earth. Some experts believe that these temperate forests harbor a diversity of species that approaches the much touted biological diversity of tropical rainforests.

As part of one of the most detailed analyses of arthropod diversity ever conducted, scientists now estimate that about 8,000 distinct species inhabit a single study site in an Oregon old-growth forest, most of them in the soil. The findings are "especially surprising because we think of that kind of diversity as being related to the tropics, not the temperate forests," said Melody Allen, executive director of the Xerces Society, an invertebrate conservation group.

But scientists say that those numbers are far less significant than still-sketchy hints of the role insects and other arthropods apparently play in the temperate forest ecosystem. "We've come to suspect that these invertebrates of the forest soil are probably the most critical factor in determining the long-term productivity of the forest," said Dr. Andrew Moldenke, an entomologist at Oregon State University in Corvallis.

In tropical rainforests, twigs, fallen leaves and dead organisms are decomposed rapidly by bacteria and fungi that thrive in the warm, wet

71

ecosystem. But in the temperate forests of the Pacific Northwest, arthropods appear to be linchpins in the decomposition process.

Billions of extremely tiny insects, mites, "microspiders" and other invertebrates serve as biological recycling engines that reduce tons of organic litter and debris, from logs to bits of moss that fall to the forest floor, into finer and finer bits. Bacteria and fungi living in the digestive tracts of the arthropods and in the soil then progressively reprocess the finely crushed, once-living tissue into basic nutrient chemicals to feed roots and, hence, the aboveground ecosystem.

New techniques for solidifying and examining soil samples offer great promise in increasing understanding of the rich ecosystems in the forest soil. Researchers caution that they still know "almost nothing" about precisely how all the thousands of arthropods interact and survive.

Taxonomists working in the region have been able to identify about 3,400 arthropod species at a single research site, the H. J. Andrews Experimental Forest in Oregon, a sort of living forest laboratory operated by the United States Forest Service. Many of those species have never before been named and described. In comparison, the combined count of all species of reptiles, birds and mammals at the site is 143.

Yet, according to Dr. John Lattin, director of the Systematic Entomology Laboratory at Oregon State University, the number of species catalogued so far probably represents less than half of the estimated species present on just the Andrews Forest site.

Simply to describe and name the yet-unnamed species will take years, in part because of the sheer numbers of arthropods in even a small area, according to Dr. Lattin. Most of the soil arthropods are exceedingly small, as tiny as one or two hundredths of an inch long. That is as small as or smaller than the period at the end of this sentence. And surveys have shown that the soil under a single square yard of forest can hold as many as 200,000 mites from a single suborder of mites, the oribatids, not to mention tens of thousands of other mites, beetles, centipedes, pseudoscorpions, springtails, "microspiders" and other creatures.

Dr. Moldenke and his students have in recent years begun studying soil and arthropod ecosystems using a technique called thin-section microscopy, originally developed by oil-exploration geologists. That approach has revealed that the very structure of temperate forest soils, and hence much of their biological and chemical activity, is determined by the dietary habits of the soil arthropods.

Thin-section microscopy is accomplished by insinuating, in a pressure chamber, epoxy into a carefully removed core of soil. Once the epoxy hardens, the now rock-like soil sample can be sliced into exceedingly thin wafers and polished smooth for examination under a microscope.

The technique preserves the soil with its parts in place, from larger bits of partly decayed plant matter to microscopic soil particles. On one such slide, Dr. Moldenke showed a visitor the image of what was clearly a needle from a coniferous tree, partly decayed, but still mostly intact.

Magnified, however, the small needle in the soil turned out to be an assemblage of thousands of infinitesimal fecal pellets arranged in almost precisely the shape of the needle. Not long after a bit of vegetation falls, millipedes descend on it, grinding it up. Chewed-up bits of vegetation pass through the insects' digestive tracts in a matter of seconds and are redeposited virtually in place as a pellet.

A closer microscopic look at each pellet reveals that each is nothing more than chopped-up bits of plant cells, reassembled into a sort of jigsaw puzzle of plant matter. These tiny clumps of cell tissue will, in turn, be eaten by other arthropods.

Deeper in the soil, the jigsaw-like cell tissues become progressively less recognizable, as successive waves of "microshredder" arthropods crush and partly digest these fecal pellets, like a series of minute millstones grinding food down to finer and finer bits.

Cell tissue cannot dissolve in water. Yet for a living ecosystem to perpetuate itself, nutrient chemicals that are locked into insoluble organic molecules in tissues of dead organisms must somehow be made soluble to be taken up by the roots of plants.

Each arthropod extracts only the whisper of nutrition from food that was once living cell matter. But in the process, each arthropod exposes more surface area to decomposer bacteria. The bacteria, in turn, biochemically process a trace more cell matter on the pellet's surface into soluble compounds, making more nutrition available to the next arthropod until, eventually, insoluble cell matter becomes soluble nutrients.

In the old-growth forest, the process is sometimes excruciatingly slow.

Soil organisms are just now completing the decomposition of some giant trees that crashed to Earth about the time Columbus sighted land.

Precisely how all these biological and chemical interactions occur, and which of the thousands of species' survival is key to the survival of others, are matters that remain poorly understood. "We've reached the point where we know just a little bit more about the fauna of the forest soil at the end of the twentieth century than was known at the beginning of the nineteenth," Dr. Moldenke said.

And researchers still don't know why there are so many invertebrate species in the forest soil in the first place. "There are still a lot of questions about why there's so much diversity," said Dr. Lattin. "But the fact that they are out there in such great numbers suggests that they play a very, very important role in the ecosystem."

Dr. Moldenke agreed. "I don't know what the implications of all that diversity are," he said. "Neither does anybody else. And that's the scary part. I guess what concerns us is that the kinds of aboveground ecosystems that most ecologists have studied in the past are a very small part of what's really out there.

"When you have an awful lot of species, it means almost by definition a great number of processes: thousands of different functions taking place. If we instead continue to manage forests on the basis that the ecosystem is much more simple than it really is, we may be setting ourselves up for a big surprise, and it may not be a nice surprise."

One potential practical benefit of all that diversity lies on the research horizon: The arthropod communities may be able to serve as an exquisitely tuned gauge of changes in the forest ecosystem.

In 1988, Dr. Moldenke began plugging data about the tens of thousands of arthropods collected from dozens of sites into a computer for statistical analysis. The results were so surprisingly consistent that he worried that the computer had been misprogrammed. Computer analysis proved that by analyzing the thousands of arthropods in a tin can full of soil from a site, a researcher could predict with accuracy the condition of the site itself.

"As a result of knowing that pattern, anyone could take a sample in the Andrews Forest and find out what time of year it was taken, whether it came from a north or south slope, what the moisture content of the soil was," Dr.

Moldenke said. "In some areas, it could tell you what kind of tree was nearby and how far away."

As a simplified example, he says, an abundance of tiny mites called Eulohmannia, which are "bright orange-yellow and look like a gasoline truck," indicate that a site is relatively dry and in a young forest. On the other hand, an abundance of Eremaeus mites, which "look like turtles with a pattern of red dots," indicates a moist site in old-growth forest.

By analyzing such characteristics among thousands of arthropod data points, a researcher may be able to monitor changes at a site brought on by, say, global warming or herbicide use.

"A tree doesn't tell you too much about what's happening," said Dr. Moldenke. "If you want to monitor change in the environment, the worst thing to look at is an organism that's centuries old. But the arthropod community allows you to look at what's happened over a different time frame, as little as a few months. And you can only do that because you have all that diversity."

—JON R. LUOMA, July 1991

Tallying Insects in a Costa Rican Forest

IN A HIGH-TECH LABORATORY in the middle of one of the world's lushest and most creature-filled rain forests, Dr. Jack Longino is watching the digitized images of ant faces flashing across a computer screen. Next to him a technician is passing a procession of beetles labeled with bar codes in front of a scanner, which beeps in recognition as though the beetles were so many cans of soup at the supermarket checkout counter.

Working out of their jungle lab at the La Selva Biological Station on the Caribbean coast of Costa Rica, these researchers are chipping away, one bug at a time, at the monumental task of inventorying every last species of creeping, crawling, flying, scuttling, teeming bug in this unthinkably diverse jungle.

While their task might seem mundane, these biologists are at the leading edge of a veritable tsunami of surveys poised to wash over the globe. A movement aimed at saving biodiversity by marketing the drugs, foods, genes and other products these wild species may carry naturally leads to the need to inventory the available goods. Biodiversity surveys like this one, which is part of Costa Rica's national inventory of biodiversity, have become all the rage.

While interest has exploded, biologists have been thrown into an intellectual and methodological vacuum. This survey of rain forest arthropods, which includes insects, spiders, mites and the like, is one of the first to offer methods for confronting these enormous stores of species. Its organizers, Dr. Longino of the Evergreen State College in Olympia, Washington, and Dr. Robert Colwell of the University of Connecticut in Storrs, are emerging as leaders in the newly born science of inventory.

"It's one of the first really big undertakings," said Dr. Michael Ivie, beetle specialist at Montana State University in Bozeman, of the project, which is known by the acronym ALAS, for Arthropods of La Selva. "They're not just going out and counting animals, they're learning new ways to determine how to do it," he said. "They're developing guidelines and goals and methodologies for doing inventories for truly diverse groups. There's really nothing else like it."

This first large-scale methodical attack on the most species-rich group in one of the most species-rich habitats in the world, meanwhile, is turning up an abundance of new and unusual creatures.

"This is very important work," said Dr. Evert E. Lindquist, a principal research scientist with Agriculture Canada, who is one of many international experts helping identify the finds. "Almost everything I'm looking at I haven't seen before. The stuff in just this one little part of a lowland rain forest in Costa Rica, it just really blows me away," he said. "I'm a world authority on a certain family of mites, but these things just don't fit into our concepts."

Sitting with Dr. Longino in his rain forest lab, one is surrounded by spiders floating in vials, wasps skewered on pins and beetles dangling from rows and rows of cardboard mounts. Outside, six- and eight-legged creatures crawl and fly everywhere, making it abundantly clear how difficult it is to catalogue this mass of steamy, teeming life.

"If you are interested in the biodiversity of a place," Dr. Longino said, "and you want an inventory, what's the most efficient way to do it, how many traps do you buy, how many people do you need? There just aren't any protocols."

So for the past three years, the ALAS team, which includes four full-time Costa Rican staff members, has made its way through this wilderness sketching its road maps as it goes. Now, 56,881 specimens and 6,151 species later, ALAS has not only gotten the hang of inventorying bugs, it has also created a kind of reference point in the otherwise murky business of species counting. When researchers do not know what to expect in a given area, they have no way of assessing how effective their methods of trapping or sampling really are.

By focusing intensely on several particular kinds of bugs—including mites, grasshoppers, katydids, weevils, scarab beetles, short-horned flies, micromoths, spiders, ants and social wasps—the researchers can begin to get a feeling for how many species there are in these groups and thus an estimate of how many more species are still uncaptured in the wild.

Dr. Longino, whose specialty is ants, says that having found some 412 species of ant at La Selva, he can then determine the effectiveness of any given trapping method by measuring how many of those 412 it can capture. After such a calibration, these standard sampling methods can be used in places where the full complement of species remains unknown, to give researchers better estimates of the total.

At the same time, ALAS is beginning to help answer some of the knottier theoretical questions that necessarily arise during such a survey. In fact, in this seemingly mundane task—catch a bug, pin it, label it, sock its data away in the computer—philosophical questions arise. What does it mean, after all, to say you are done?

"You just keep accumulating species," said Dr. Longino. "They keep coming in." And even after a survey has gone on seemingly forever, he said, "there are still species for which you have only a single specimen."

While in the past researchers had to rely on intuition to decide when they were done, Dr. Colwell and Dr. Jonathan Coddington, a spider specialist at the National Museum of Natural History at the Smithsonian Institution, suggest that simple statistics can help researchers decide when to hang up their nets. In one method, they found that by comparing the number of very rare species in an inventory with the total, researchers can begin to get a reasonable estimate of just how close they are to the task's end.

With such quantitative methods in hand, researchers say scientists can begin to come closer not only to effective inventorying but also to knowing what scientists call the elusive grail number—the total count of species in the world.

Much of the enthusiasm for inventories and their potential for making a viable industry out of conserving species can be credited to Dr. Dan Janzen, professor of biology at the University of Pennsylvania in Philadelphia and a tireless promoter of biodiversity. Dr. Janzen is also a technical advisor to Costa Rica's National Biodiversity Institute, with whom ALAS, itself a part of the Organization for Tropical Studies, works in close collaboration. Put simply, Dr. Janzen views conserved areas as national greenhouses full of products that make them worth saving.

"I think there's a great deal of promise," he said, expressing an ever more common sentiment among researchers. "Today so few genes and crop plants and organisms are currently used in comparison with the very large quantity represented in a large tropical wild area. As I see it, it's as though we had the Library of Congress and we'd read ten books out of it."

In many ways researchers like those at ALAS are direct intellectual descendants of the nineteenth-century tropical explorers like Charles Darwin and Henry Walter Bates.

Heirs to a long and venerable tradition of exotic creature-nabbing and naming, today's high-tech explorers are continuing the long tromp through the muddy jungle. Poring over the tedium of ant after ant after ant, they keep on in the hopes of finishing the job of making sense of the glorious mess of tropical nature.

—CAROL KAESUK YOON, July 1995

In Spring, Nature's Cycle
Brings a Dead Tree to Life

IF YOU DON'T BELIEVE there's life after death, look closer some spring day at a dead tree lying on the forest floor. Chances are, if it has been there for a while, it is teeming with more life now, after death, than when it was standing erect lifting its leafy arms to pray.

Though it lacks the spring finery that inspires poets and lovers, a leafless tree is often more valuable to its forest dead than alive, say ecologists working in the old-growth forests of the Pacific Northwest. This fact, they say, has been largely ignored by wood-hungry forest managers in most of the United States and Europe, where overzealous harvesting of "dead wood" has depleted forests and rendered them highly susceptible to environmental stresses like acid rain.

"Rotten wood was once considered just a fire hazard, a waste, an impediment to travel," remarked Dr. Michael Amaranthus, a soil scientist with the United States Forest Service in Grants Pass, Oregon. "More and more we are seeing it as an essential part of the forest system, crucial to its long-term productivity. It provides a reservoir of moisture and nutrients and a variety of habitats and food resources for a wide diversity of organisms. Our understanding of the importance of dead wood has increased a lot in the last ten years."

When nature cries "timber," countless unseen denizens of the forest rush to take up lodging in the fallen tree. Dead trees serve as warehouses and even factories for essential nutrients that enrich the soil and foster new growth. They store carbon, thus curbing atmospheric carbon dioxide and the pace of global warming. They hold volumes of water that sustain growing trees during droughts. And they serve as nurseries for new plant life, providing cozy niches where seeds can gain a firm roothold and outgrow

Birth, Death and Renewal in a Tree
When a tree dies, life does not end. In a series of steps (clockwise from here), insects, plants and fungi colonize the decaying tree, and it evolves into a new habitat, teeming with life.

3. Nutrition from fungi and bacteria
Microorganisms proliferate and help decay the wood. Fungi and bacteria are key sources of nutrients, especially nitrogen, for plants and animals that thrive in dead trees.

2. Mosses and lichens
Mosses and lichens become established on the surface and capture nutrients in rainwater that has passed through the forest canopy.

1. Opening the tree to the external world
Bark- and wood-boring beetles, carpenter ants, and mites initiate the invasion by channeling through the wood. Fungi, insects and roots follow these paths.

8. Regenerating Soil
Nutrients return to soil in decaying plant and animal material, completing the cycle of life, death and renewal.

4. Grazing on fungi and microbes
The fungi are foodstuffs for many invertebrates, like beetles and mites.

Patricia J. Wynne

7. Small mammals arrive
Voles and shrews burrow in soft wood and feed on mushrooms and truffles.

5. Grazing on the grazers
Larger arthrodpods like spiders feast on invertebrates that dine on fungi and microbes.

6. Larger plants colonize
Roots of seedling trees and shrubs, like hemlocks, penetrate log crevices, aided by a symbiotic relationship with the fungi.

other seedlings struggling to capture the light that penetrates where the tree once stood.

The trunk of a dead tree is consumed by a varied succession of microbes, plants and animals, which help to replenish the soil as they break down the wood. A result, say the two forest ecologists, Chris Maser and James M. Trappe, is "an accumulation of life and nutrients that is greater than the sum of its original parts."

"In a forest where the trees are repeatedly cut and removed, the soil becomes depleted, the structures deteriorate and the forest loses its resilience for coping with stress," said Dr. Trappe, a forest mycologist at Oregon State University in Corvallis. This has already happened in Germany, where the forests are being severely damaged by air pollution and acid rain, he said. "And Germany is the country whose concept of intensive forest management served as a model for our own," he noted.

Fallen trees help to preserve the forest by stemming the erosion of soil from wooded slopes and diverting streams that in straight courses might gouge out soil. In fresh waterways, fallen trees trap nutrient-rich sediments and create pools where fish can spawn and fry develop.

Beyond the forest, dead trees help stabilize beaches and create habitats for wildlife in estuaries and salt marshes. Logs that reach the open sea serve as a major source of carbon and other foodstuffs for marine life.

"Unfortunately, very little of this is now happening because the oceans are being deprived of this resource," said Mr. Maser, an author and consultant living in Las Vegas, Nevada. "We are beginning to starve the oceans as well as the soil because we are not reinvesting the biological capital nature provides into the forest, ocean, air or land."

"The function of dead trees in the ecosystem has rarely received the consideration that it deserves," says Dr. Jerry F. Franklin, an ecosystem analyst at the University of Washington's College of Forest Resources in Seattle. "At the time a tree dies, it has only partially fulfilled its potential ecological function. In its dead form, a tree continues to play numerous roles as it influences surrounding organisms. The woody structure may remain for centuries and influence habitat conditions for millennia."

So, these forest scientists urge, woodsman, woodsman, spare thine ax for fallen as well as standing trees. Think twice before hacking up and carting off those logs dead in name only and dooming them to a brief and limited life as firewood.

Once a tree falls, it passes through five distinct phases of decay, Maser and Trappe wrote in their technical review, "The Seen and Unseen World of the Fallen Tree," published in the 1980s. At each stage, the tree supports new life for which it is the sole or principal habitat.

At stage 1 are newly fallen trees with intact bark, a condition soon to be changed as bark and wood-boring beetles tunnel through. These brazen beetles blithely disregard the chemical and mechanical defenses of the conifer's bark that discourage most insect predators. The first beetles create channels for their successors. The beetles also carry in fungi and bacteria that provide food and essential nitrogen for future invaders.

At stage 2, trees still retain bark but as the beetles feast away, the nutritious growing layer of inner bark and the nearby phloem, which transported sugars, become spongy. These tissues are likely to be eaten in a few years. Next in line is the sapwood, which in the living tree housed the water-carrying structures called xylem.

By stage 3 the bark sloughs off. Roots from sprouting seeds now invade the sapwood, and the trunk begins to break into large, solid pieces. In a fallen Douglas fir, the sapwood succumbs to insects and fungi in 10 to 20 years, Dr. Trappe said, although the bark of this tree "probably hangs around for centuries."

At stage 4 the heartwood, composed of the dead xylem that forms the bulk of the tree trunk, is all that remains. It now breaks apart into soft blocks as roots invade this dense, highly resistant and not very nutritious wood. This is the stage, the longest in the decay process, that hosts the most diverse array of wildlife, including mites, centipedes and snails, as well as salamanders, shrews and voles.

Finally, in stage 5, the tree is no more than a soft, powdery mass. Ashes to ashes, dust to dust, soil to soil.

Stocked with nutrients, a fallen tree supports more life than when it was alive. Invading fungi ooze out enzymes that liberate the tree's nitrogen for use by other organisms. More nitrogen is provided by bacteria that extract it from the air. The tiny organisms that inhabit the log fertilize it with their excrement. Leaf litter and rainwater laden with nutrients and lichens from the forest canopy fall on the dead tree, adding further enrichment.

Carpenter ants are most active in stage 2. Their catholic diet includes butterflies and the honeydew of aphids. Nesting in fallen logs, they carry nutrients into the tree from the outside. Termites take over late in stage 2, importing in their wood-chomping bodies both protozoa that digest cellulose and bacteria that capture atmospheric nitrogen. By the time a termite colony is ready to move on, it has created a labyrinth of passageways in the tree that can be used by other animals and by the roots of invading plants.

As logs reach stage 3, their bark and sapwood is sloughed off and plants have taken root. The logs become ready for occupation by a wide range of animals. As Mr. Maser and Dr. Trappe wrote about the trees when they reach stage 4, "Various mites, insects, slugs and snails feed on the higher plants that become established on the rotten wood. These plants also provide cover for the animals, as do the lichens, mosses and liverworts that colonize fallen trees."

In this microenvironment, mites thrive on the dead plant and animal matter that accumulates on fallen trees. The skeletons of dead mites, in turn, serve as incubators for fungal spores, and the fungi provide sustenance for other invading plants and animals.

The folding-door spider is among the many arthropods that thrive in these conditions. It constructs a silky tube in one of the many cracks in the outer layer of a fallen tree that has reached stage 3 or 4 of its decay. The outer edges of the tube are pulled inward to form a slitted cover and the spider waits on the inside for the arrival of suitable prey, which are abundant in the decaying wood.

Among the ecologically important denizens of fallen Douglas fir is the California red-backed vole. The rodent eats mostly fungi and lichens but has a particular passion for truffles, Mr. Maser has shown. The vole then disperses the spores of the truffle, inoculating decaying trees with this valued foodstuff. This benefits other truffle-eaters, including the squirrels and mice that are the principal foodstuffs of the spotted owl and other carnivores.

"The spotted owl debate is not a case of owls versus people," Dr. Trappe said. "It's a question of whether we want the diversity of organisms that the natural forest provides, or in its place a monoculture in which many organisms will disappear, not just the spotted owl."

If Dr. Mark E. Harmon, a forest ecologist at Oregon State University, has his way, dead trees as well as living forests will become valued as critical elements in containing global warming. When a tree is cut and processed into paper or a fallen tree turned into firewood, carbon dioxide is ultimately released into the atmosphere. "But a dead tree left on the forest floor holds on to its carbon for decades, even centuries," he explained.

Dr. Harmon is directing a project whose lofty time horizon rivals that of the Earth-made plaque sent aboard the spacecraft *Pioneer 10* to Jupiter and beyond. More than 500 logs of four different species have been placed throughout the H. J. Andrews Experimental Forest outside Eugene, and their patterns of decomposition are to be studied over the 200 years they will take to decay. Biologists will monitor the insects and microorganisms that colonize the logs, the small plants and large trees that become established on them and the birds, reptiles and mammals that use them as dwellings and food sources.

In a parallel experiment on two sides of the Cascades, 800 large trees were felled in 1987 and 1988 and placed in streams. Dr. James Sedell, an aquatic biologist at Oregon State University, said the project had already restored habitats for juvenile coho salmon and steelhead trout.

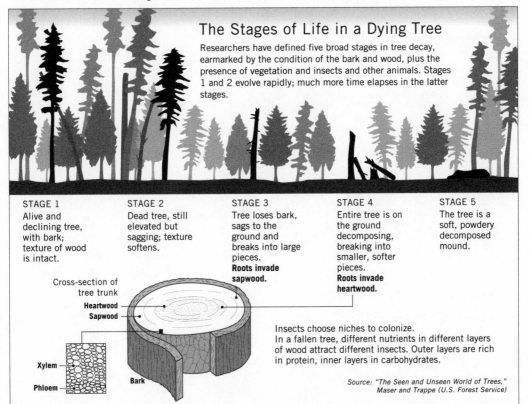

The Stages of Life in a Dying Tree

Researchers have defined five broad stages in tree decay, earmarked by the condition of the bark and wood, plus the presence of vegetation and insects and other animals. Stages 1 and 2 evolve rapidly; much more time elapses in the latter stages.

STAGE 1
Alive and declining tree, with bark; texture of wood is intact.

STAGE 2
Dead tree, still elevated but sagging; texture softens.

STAGE 3
Tree loses bark, sags to the ground and breaks into large pieces.
Roots invade sapwood.

STAGE 4
Entire tree is on the ground decomposing, breaking into smaller, softer pieces.
Roots invade heartwood.

STAGE 5
The tree is a soft, powdery decomposed mound.

Cross-section of tree trunk
Heartwood
Sapwood

Xylem
Phloem
Bark

Insects choose niches to colonize. In a fallen tree, different nutrients in different layers of wood attract different insects. Outer layers are rich in protein, inner layers in carbohydrates.

Source: "The Seen and Unseen World of Trees," Maser and Trappe (U.S. Forest Service)

"When a large log falls in a stream, the current scours out a pool around it and other wood gets trapped to form a debris jam," Dr. Sedell said. Fish then go into the pool, which serves as a safe harbor during winter floods and a secure habitat in summer droughts, he explained. The next step is to see if more fish leave the stream and grow up in the sea.

"I'm optimistic," the biologist remarked. "Worldwide there's been much more interest in the role of wood in rivers and streams. The Forest Service and several states have begun to recognize that on forested land they need to allow big fallen logs to remain in streams to protect the fish resources."

Now, he and other scientists say, the question on land and water is: How much dead wood must be kept to bring back the many habitats needed to sustain the diversity of life on earth?

—JANE E. BRODY, March 1992

2

ATTACK AND DEFENSE

For insects, too, there is no free lunch. Almost everything in an insect's world comes with difficulty. That caterpillar munching happily on a leaf has evolved the chemical means to detoxify the poisons that plants release when they sense they are being eaten. That gliding butterfly has evaded the insidious traps of spiders' webs and the aerial attacks of hungry birds.

Plants and insects have been fighting each other for so many eons that they have complex interactions with one another. The oak tree fills its leaves with tannin to discourage diners; gypsy moth caterpillars have learned not only how to eat the leaves but also how to use the tannin as chemical protection against viruses.

Parasitic wasps that lay their eggs inside caterpillars find their prey by sensing the chemicals released by plants in response to being lunched on. The plants and the wasps seem to have made common cause against the caterpillars, since some plants release their chemicals at just the time that the wasps are out hunting.

When it is so hard to make a living, there are high rewards for finding a niche all to one's own. Insects have learned how to adapt to almost every extreme of habitat. The desert ants of the Sahara are a fine example of living at the brink of the possible. The sands are so hot that the ants must sprint across them, climbing grass stems every so often to cool off.

As the following chapters attest, insects work hard for a living.

Crafty Signs Spun in Web
Say to Prey, "Open Sky"

THE ROOM BECKONS warmly and widely, lighted high overhead by glad-
dening bulbs that recapitulate the brilliance and spectral range of the sun.
But as you step into this parlor of a laboratory just try not to jump, or at the
very least stiffen.

In every corner, under every surface, dangling into one's hair, brush-
ing against one's shoulder, are large, dainty, lacy, spiraling, glittering spiderwebs. And tending each of these webs is a large, undainty, big-bellied,
generously appendaged spider. There are dozens and dozens of spiders,
some of them yellow, some brownish black, some pale caramel. And all of
them are weirdly frozen in place because—what do you know!—spiders
really are more scared of humans than most humans are of them.

"A spider's first impulse when something larger comes along is to stop
moving and hope it goes away," said Dr. Catherine L. Craig, an evolutionary ecologist at Yale University, who is studying the evolution of spiderwebs.
She taps on a web to prod the little architect from its stupor. It skitters briefly
and freezes again.

"There are cultural biases against spiders," Dr. Craig said, with some
understatement. "Most people look at my roomful of hanging spiders and
it's a nightmare for them. I had one student who came to my lab and vol-
unteered to feed the spiders, but when she saw the spider room, her face
contorted like this"—Dr. Craig gives her freckled face a Munchian twist—
"and she said, 'EEEUUUWWW!'"

No such squeamishness for Dr. Craig. "I can't think of any place I'd rather
be," she said, "than sweating in the sun in Panama playing with spiders."

Dr. Craig divides her time between field studies at the Smithsonian
Tropical Research Institute on Barro Colorado Island in Panama and labo-

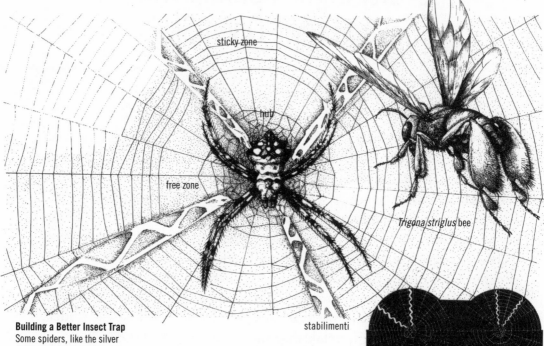

sticky zone

hub

free zone

Trigona striglus bee

stabilimenti

Building a Better Insect Trap
Some spiders, like the silver *Argiope* above, weave struts called stabilimenti into their webs to entice flying insects by reflecting ultraviolet light. Part of the spider's body is also reflective in the same ultraviolet range. The rest of the sticky web has markedly reduced reflectivity, making it invisible to prey like this 0.4-inch stingless bee.

Elegant and Deceptive Variations
By shifting the pattern of the stabilimenti from day to day, the spider seems to interfere with the bees' ability to learn from their mistakes. Here are four variations on a two-bar embellishment of a web.

Michael Rothman

ratory work at Yale with her in-house collection of tropical species of spiders that weave orb webs, including *Argiope argentata*—a relative of the common garden spider found in the Eastern United States—and *Nephila clavipes*. These are the spiders that generate the cobwebs of Halloween fame, as well as the less-familiar ladder webs, funnel webs, hanging webs and other lattices plain or fancy. Their webs are called orbs in the old-fashioned sense of the word, meaning circular.

To Dr. Craig, a spiderweb is not a passive structure or a simple sieve that catches insects that blindly fly into it, as had long been believed. Instead, she views the web as among the spider's most dynamic and responsive traits, a cunning weapon designed to lure prey by exploiting an insect's fundamental need for food, flowers and open spaces.

She proposes that spiders incorporate into their webs visual signals like attractive zigzag designs and faux floral colors that are irresistible to a wide range of insects. And because these signals are the cues insects use to forage for food, the creatures are not likely to evolve a means of detecting and avoiding the trickster webs without simultaneously jeopardizing their skill for finding a meal.

In an outpouring of reports put together over the last several months, Dr. Craig offers evidence for a genuine spider revolution at an unknown point in the past, resulting from minor modifications in the silk proteins of which webs are built. She focuses on the subtle details of the webs, trying to understand how the visual and mechanical properties of the silks that make up the webs may have changed over evolutionary time and contributed to the extraordinary success and explosive diversification of orb-weaving spiders.

The molecular changes improved the strength, elasticity and versatility of web silks and allowed the ancestral orb-weaving arachnids to emerge from their obscurity in the dim forest understory and begin laying traps in the open sun and other theretofore forbidden habitats, where whole new classes of prey became available for their dining pleasure. As a result of the evolution of a refined type of silk, she argues, the number of orb-weaving species increased by a stupendous 37-fold over the more primitive orb spinners that predated them. They have branched off into at least 10,000 different species, which means they represent almost a third of all spiders described to date.

The results are presented in three papers published in the journals *Animal Behaviour; Evolution,* and *Behavioral Ecology and Sociobiology.* In the studies, Dr. Craig and her colleagues combine detailed spectral analysis of spider silk with field studies of how stingless bees—a major source of prey for orb weavers—perceive and respond to variations in spiderweb design.

The results have far-reaching implications for understanding essential questions of ecology and evolution, among them the nuances of predator-prey interactions, and the mechanism through which new species arise. Dr. Craig sees the spiderweb as a beautiful means for weaving a molecular approach to biology with a more holistic view of animal behavior. "It's a way to tie together evolutionary studies of macroprocess like foraging behavior with considerations at the level of the genes," Dr. Craig said, in this case the genes that direct the spider's all-important silk production.

Dr. Craig also believes that by understanding the complex spectral features of a spiderweb, scientists can get a handle on how insects see their world.

"Her work has been very interesting for a lot of us," said Dr. George W. Uetz, a professor of biology at the University of Cincinnati who has studied spider foraging behavior. "It's made us look at insect-spiderweb interactions differently."

Dr. Uetz said that his earlier research had suggested that spiderwebs were passive filters for collecting insects, rather than real attractants for prey. "Her work has overturned that whole set of ideas," he said. "You might think I'd be depressed about it, but I'm not."

Dr. Uetz added that a more sophisticated knowledge of spider foraging strategies would help him in his own investigations of an unusual type of tropical spider in Mexico, which gathers in groups of up to 165,000 spiders per colony, building enormous communal webs across trees or power lines, and interacting socially, a rarity among spiders.

Dr. Craig began her study of spiderwebs nearly a decade ago, inspired by the intellectual possibilities of the subject rather than any sort of lifelong love of the beasts. "The last time I stepped on a spider," she said, "was the day before I began working on them."

She saw the web as a relatively neglected area of research, one that had been subjected to rather silly and uninformative manipulations—like giving spiders hallucinogenic drugs and observing the mess they made of their webs.

Her first step was to investigate the common assumption that webs were simple colanders, invisible to the luckless insects that barreled into them. Such a supposition meant that interactions between web-building spiders and insects were passive and indirect, very different from the active encounter between, say, a lion and the gazelle it covets.

Dr. Craig soon proved these assumptions wrong. Through a series of experiments in which she examined and quantified insect responses to webs in different lighting conditions, she found that the potential prey could indeed see the silk meshes under certain circumstances. And often, when they got close enough to the web to make it out in all its filamentous horror, they made unmistakable efforts to fly around the trap.

Those findings prompted her to wonder whether spiders might seek to counteract that repulsion and, if so, how they did it. She first showed that many web silks have a noteworthy feature: They reflect light in the ultraviolet range of the spectrum. This property may serve to attract insects, which unlike humans can see ultraviolet light and are drawn to it from afar as part of a so-called open-space response: Insects need open space to help them navigate, and because ultraviolet light can come only from the sun or the sky, a bit of it glittering is like a billboard proclaiming free range ahead.

But in her most recent work, some of it done with Dr. Gary D. Bernard of the University of Washington and Dr. Jonathan A. Coddington at the Smithsonian Institution, Dr. Craig has proved that ultraviolet shimmering is only one part of the orb weaver's story. As it turns out, this simple reflectance feature is a primitive characteristic of spiderwebs, and the more advanced orb weavers have come up with clever variations on the theme.

In the new experiments, the scientists have focused on two big web weavers, *Argiope* and *Nephila*. Both spiders hunt in broad daylight, building new webs each morning to catch fast-flying, keen-eyed meaty creatures like bees, and consuming the silk traps each night to use their protein. The spiders' webs display markedly different optical properties, yet Dr. Craig has shown that both approaches effectively seduce—and confuse—the prey.

With artful flair, the *Argiope* spiders decorate their webs, placing thick strands of silk in the middle to create zigzags or crosshatch patterns. The decorations appear white to us, but the scientists have shown that the adornments reflect ultraviolet light brightly, just as do the web silks of primitive spiders. At the same time, the silks making up the rest of the web, all the circling and radiating strands designed to catch and hold prey, show a marked reduction in their ultraviolet-reflecting properties.

Dr. Craig suggests that the decoration entices prey, not by stimulating a generalized open-space response, but by resembling blossoming grasses or nectar guides on flower petals, which also glow with ultraviolet light in a more benign effort to invite pollinators. Because the surrounding web is essentially invisible to the insect, the prey would be drawn toward the decoration, then get tangled up in the unseen catching threads around it.

Significantly, the spiders vary the placement of the decorations from one day to the next. In other new experiments, Dr. Craig has demonstrated that the variation is important to the spiders' foraging success. It turns out that many bees intercepted by webs manage to shake themselves free. By shifting the pattern of the central embellishment from day to day, the spider seems to interfere with the bees' ability to learn from their mistakes.

That is good news for the spider, not solely because the next time the repeat visitor may get stuck for good and thereby serve as a substantial meal, but also because getting caught even temporarily benefits the spider: She appears to consume as a vegetarian side dish the pollen packets that escaping bees leave behind as they frantically disengage from the web.

For its part, the giant *Nephila* uses pigments to make its web appealing to prey. The scientists have learned that the spider will incorporate more or less yellow pigment into its silk strands depending on how bright the sun is over a period of several days: The more sunshine, the more yellow the web becomes with each passing day. Dr. Craig proposes that yellow is a brilliant choice of color from a strategic perspective.

"Yellow is a very generalized visual signal that both herbivorous and pollinating insects associate with flowers and new leaf material," she said. If the spiderweb likewise has the color of the most abundant food resource, she added, the insects may have a much harder time evolving a mechanism to avoid it.

In field experiments with bees trained to forage for sugar water at the same spot several times a day, Dr. Craig and her assistants presented the bees with webs, hung on a small hoop and artificially colored blue-violet, green, yellow or unpigmented. With most of the webs, the bees quickly learned to fly up and over the hoops to get to the sugary reward. But in the case of the yellow-tinted web, the bees were confused, either repeatedly getting caught, or flying up to it and away.

Putting together the various strands of evidence, the scientists propose that the suppleness of their silks distinguishes the advanced—or as scientists prefer to call them, derived—orb-weaving spiders, like *Argiope,* from their primitive cousins, a small group of spiders called the deinopoids. The deinopoid spiders, the scientists suggest, are comparatively older and less varied in their web-making techniques.

Their silk uniformly reflects ultraviolet light. It is relatively weak and rigid. As a result, the deinopoids are limited to building webs at night or in dim light, where they may be hazily attractive to insects but not seen in any sharp, repellant particulars. They can catch only slow or small insects.

Not surprisingly, then, the primitive orb-weaving spiders count a mere 300 species or so in their group.

By contrast, the advanced orb weavers, called the araneoids, possess silk of great optical versatility and superior strength and elasticity. The spiders have moved everywhere, learned to do everything and made practically the entire insect kingdom their oyster.

Dr. Craig next plans to investigate the molecular and genetic differences between the primitive and newer types of silks, but already it is clear that the triumph of the great orb-weaving spiders hangs by a few slender threads.

—NATALIE ANGIER, April 1994

Nibbled Plants Launch Active Attacks

FOR MORE THAN 100 MILLION YEARS the battle has endured, and it rages still in every pine grove, meadow and planted field. It is the truceless war between plants and the predators that feed on them.

Besieged by armies of voracious creatures but unable to run away, plants over the eons have evolved cunning defenses that include deadly poisons, oozings of toxic glue and hidden drugs that give leaf-eaters serious indigestion.

These defenses are of great interest to biologists because they appear to work as natural pest controls, restricting most insects to the few plants whose defenses they have somehow been able to overcome or withstand.

In the latest twist in the study of plant defenses, scientists are finding that many species have evolved a sophisticated and prudent strategy, that of holding their best weaponry in reserve. Many plants make toxins in their leaves, but in addition to these passive defenses, several species wait until a predator actually starts munching before they unleash their most noxious washes of chemicals. By learning how to manipulate these natural defenses, agricultural researchers hope to develop potent alternatives to pesticides. Several of the developments in this rapidly unfolding field of research are described in *Phytochemical Induction by Herbivores,* published by John Wiley & Sons.

"People just haven't really viewed plants as the dynamic, aggressive pugilistic little beasts that they are," said Dr. Ian Baldwin, a biologist at the State University of New York at Buffalo. "There's a war going on out there."

One of the most violent battlegrounds is the pine forests of the Western United States, where mountain pine beetles struggle against lodgepole and ponderosa pines.

"It's a life or death situation," said Dr. Kenneth Raffa, an entomology professor at the University of Wisconsin. "In order for the insect to reproduce, it has to kill the tree. Either it kills the tree or the tree survives by killing or repelling the insect."

It all begins with the arrival of a single, small, dark brown female beetle no bigger than a grain of rice. She bores an entry hole through the bark and into the tree.

As soon as the first beetle has bored in, the pine begins its own counterattack. It starts killing off the cells around the wounded area and flooding the invading beetle with sticky resins that gum up its paths and clog its entry hole.

As the resins pour in, the beetles begin shoveling them out, an activity that may continue for days or weeks before they can safely chew out chambers clean enough in which to lay their eggs. As the mere touch of a bleeding pine tree will show, the resins are remarkably sticky and difficult to wipe off. Yet these beetles can scramble across it and shovel it away, somehow immune to the gooey mess.

The beetles can even walk through the sticky chemicals used as commercial insect traps, Dr. Raffa said. Chipmunks can get stuck in them, "but the beetles walk right through it," he said.

The pine tree's resin is laden with special chemicals, known as terpenes, that poison the air and the beetles' newly dug brood chambers. The terpenes are what gives pine forests their characteristic fresh smell.

While the tree is trying to isolate the beetle in a mass of dead cells and harmful glues, the female beetle in a counterattack tries to call in support troops to help her take her gigantic enemy down.

For Insects, the Buzz is Chemical

When larval caterpillars begin munching on the leaves of, say, a corn stalk, their saliva somehow activates the plant to begin emitting chemicals like terpenes, the active ingredient in turpentine. The parasitic wasps pick up this signal and find their unwitting larval hosts. Cutting the leaves with a scissor will not do the trick; only the caterpillar's mouthparts seem able to set off the alerting terpene mix.

Of particular interest to the scientists, an infested plant appears to recognize the value of the parasitic wasps that will kill its enemy. The plant will not begin emitting the chemical signal at night, when the caterpillars start feeding, but rather delays the release until the following morning, when the parasitic wasps are likely to be out hunting. What is more, the damaged stalk releases the giveaway compound in sharp peaks, each corresponding to the moment of prime wasp hunting hours, as though sending out an SOS.

—Natalie Angier

Plowing through the resins, she begins eating some and converting their terpenes into a special, very volatile perfume. The perfume so excites other mountain pine beetles that thousands can soon descend on a single tree and start drilling into it.

"It all happens so quickly you can hear it," Dr. Raffa said. "It sounds like somebody ripping cardboard and the whole thing could be over in two or three days."

The beetles drill and the tree bleeds out its poisons. Usually the trees win and the beetles are killed or forced to move onward, but sometimes the amassed armies of boring beetles overcome the tree's defenses, and then they make their homes in the stricken trunk and raise their young.

Potato and tomato plants opt instead for sneakier methods of attack than the pine. They fight back by giving their enemies indigestion.

Dr. Clarence Ryan and his colleagues at Washington State University are piecing together in minute molecular detail the puzzle of how potatoes and tomatoes stop their enemies from digesting their food.

They have found that when a caterpillar chews on a potato leaf, bits of broken cell wall and other chemical signals begin flowing through the plant, alerting it to trouble. In response, the plant begins making chemicals that inhibit those that help the insect digest. The caterpillar continues to eat, but the digestion-inhibiting molecules deprive it of vital nutrients and retard its growth. This gives the caterpillar's predators longer to attack it before it pupates.

Other plants go straight for the kill, simply flooding their leaves with deadly poisons.

The coyote tobacco plant, long smoked by the Anasazi Indians, is as well defended as a "little tank," Dr. Baldwin said. When attacked, it pumps its leaves full of nicotine, increasing the amount of the poison tenfold. According to Dr. Baldwin, a single leaf weighing one thirtieth of an ounce can kill 10 rats.

"Even well-adapted plant-eaters get sick on this," he said. "Hornworms eat it, ground squirrels, cows, much to their gastronomic distress. I've seen jackrabbits eat this stuff and get diarrhea."

Nicotine belongs to a family of plant chemicals called alkaloids. Other alkaloids thought to protect plants from grazers include cocaine, caffeine and many other chemicals used by humans as drugs.

Like coyote tobacco, the tall umbels of wild parsnip flowers that add sunny color to fields and roadsides are dangerous reservoirs of poison. When under siege, these common weeds employ a toxin, called a furanocoumarin, that can get into cells and hook together strands of DNA, the genetic material, causing serious illness.

Wild parsnips are "toxic to most DNA-based life, which is basically everything," said Dr. May R. Berenbaum, a professor of entomology at the University of Illinois at Urbana-Champaign. "It's on poison plant lists. If you touch it in the presence of light it causes a painful itchy rash that's often misdiagnosed as poison ivy."

Like coyote tobacco, wild parsnips when they are losing leaves can tell whether it is from animals munching or just the battering of wind or rain. They release their poisons only when under attack by a grazing animal.

"It makes perfect sense," Dr. Baldwin said. "An alkaloid response to breakage from wind or a branch falling would be totally inappropriate, whereas it would very appropriate for a caterpillar."

Researchers suspect that something in the saliva of caterpillars tells the plant that it is being attacked by an animal rather than a gust of wind or a pair of scissors. "Even as we speak, people all over the country are collecting caterpillar spit" to try and answer the question, Dr. Berenbaum said.

For every clever defense that plants devise, some animal seems to be able to come up with a countermeasure. Parsnip webworms can detoxify the parsnips' DNA-linking chemicals. Another animal, so far unidentified, has learned to get around the coyote tobacco plant's defenses by chewing a girdle around its stem, destroying the channel down which the plant sends its emergency mobilization message when under attack. Having cut off the plant's lines of communication, the animal can eat its fill of the still tasty plant.

Even more cunning are the animals that have found ways to convert the poisons directed against them into assets. Dr. Jack Schultz, an entomology professor at Pennsylvania State University has found that gypsy moth caterpillars feeding on oak leaves can use toxic defense chemicals known as tannins to fight off marauding viruses.

The gypsy moth caterpillars actually prefer leaves with tannin, even though they grow faster on a tannin-free diet. When offered an artificial caterpillar diet supplemented with viruses, they reject it as though they can somehow sense the viruses. Yet when tannin is added, they will accept the food even if it is laden with viruses.

Dr. Pedro Barbosa, a professor at the University of Maryland, has found another example of plants' defenses being appropriated. The alkaloids in tobacco and tomato plants, when eaten by caterpillars, will kill tiny wasp larvae, parasites that feed on the caterpillars' internal tissues.

Biologists hope that ways can be found to trigger these active or inducible defenses in crop plants. Researchers at the University of California at Davis are working on "vaccinations" for plants. Dr. Rick Karban and his colleagues expose grape plants to harmless mites, prompting the plants to turn on their defenses. When the real pest mites arrive later in the growing season, the plants are braced for attack. Successfully defended, the grapes end up growing sweeter faster, making them more valuable at harvest time.

Dr. Karban said he hoped the vaccination approach would help farmers "get off the pesticide treadmill."

—CAROL KAESUK YOON, June 1992

Beetle Strategy Outwits Toxin-Spewing Plant

DEEP IN THE HEART of the Tehuacán Desert valley in Mexico, a researcher has discovered a plant bearing leaves that can shoot a toxic jet of chemicals a distance of two yards. Heavily armed, the plant would appear to be safe from all insect attackers. But a study has found that one beetle can chew its way past the plant's defenses by severing the canals that deliver the toxins to the leaves.

Until recently researchers thought insects simply had to swallow and somehow contend with poisons lacing a plant's leafy tissues. But these bugs, they are discovering, have come up with more ingenious solutions. Scientists say the new study, published in the journal *Ecology,* provides the latest addition to a growing list of clever chewing counterploys that insects use to deactivate or avoid plants' noxious defenses.

Calling the new study "outstanding," Dr. David Dussourd, a behavioral ecologist at the University of Central Arkansas in Conway, explained that vein-cutting is just one method for preparing a safe meal. Another is trenching, cutting a swath all the way across a leaf to halt the flow of chemicals past it and simply eating daintily between the canals of resins, glues or toxins.

Dr. Robert Denno, an insect ecologist at the University of Maryland, said: "Historically scientists studying insect adaptations to plant defenses have focused on detoxification and excretion. What's unique about these strategies is that they're actually behavioral methods of circumventing plant defenses."

Dr. Judith X. Becerra, a postdoctoral fellow at the University of Colorado in Boulder, said she was first attracted to the desert shrub, known as a bursera, by the heady smell of its resins. Stored in canals running through the plant's leaves, these are the resins used in copal, the ceremonial incense of the ancient Mayans. Hoping to enjoy the pine-like scents herself, Dr. Becerra broke some bursera leaves, inadvertently discovering their ability to squirt.

But the resins are considerably less enticing to the young larvae of the plant's enemies, flea beetles, so named for their jumping abilities. Biting into the leaves and through tributaries of toxic networks, the larvae are bathed in the plant's sticky, odoriferous surprise. As resin pours or shoots out, it glues up mouthparts and can even entomb larvae. Only after a larva survives several such insults from a plant does it begin its counterattack, chewing methodically through the resin-transporting veins on the next leaf it eats.

After seeing the effectiveness of the insects' vein-cutting and trenching, scientists assumed such specialist insects had won out. Researchers figured the plants' defenses were only effective against species unable to deactivate the canals. But Dr. Becerra found that though the flea beetle's handiwork does the trick, it comes at a high cost.

She found that a small leaf that takes a beetle only a few minutes to eat can take an hour and a half of careful vein-severing to prepare. Researchers say such investments of time make the food costly and dangerous. The longer it takes to cut the vein, the longer it takes to get the food necessary to mature to adulthood and the longer the developing larva is exposed to predators and parasites. So while the plant may lose the battle when a beetle severs its leaf, it may end by winning the war as its insect attackers struggle to remain safe and to get enough to eat.

—Carol Kaesuk Yoon, December 1994

Life at the Extremes:
Ants Defy Desert Heat

BY MIDDAY the temperature at the desert's surface has climbed to 140 degrees. As far as the eye can see to the shimmering horizon at the desert's edges, there is not even the tiniest patch of shade. A few animals that have strayed too far from their underground shelters collapse from exhaustion under the Saharan sun.

Just as every other creature has taken refuge from the impossible heat, hundreds of Saharan silver ants burst out of their nest hole into the blazing noonday sun. It's time to go a-hunting.

Powerfully resilient to the heat, these ants can withstand higher ground temperatures than any other animal and scavenge the desert when all other animals have been driven to escape the heat.

The ants' particular enemy is a lizard that builds its burrow near their nest holes. Their life aboveground is thus bounded by a narrow range of extreme temperature, searing enough to drive the lurking lizards into their burrows but not so fierce that the ants themselves drop from exhaustion.

From deserts to glaciers, animals seem to have invented ways to survive in even the harshest corners of the world. There are bacteria that are able to thrive in boiling hot springs as a result of heat-resistant proteins. Antarctic fish and insects pump their cold bodies full of antifreeze to resist becoming frozen solid. Strange worms and bugs living by deep sea vents survive on toxic chemicals that spew from openings in the sea floor.

Dr. Edward O. Wilson, a biologist at Harvard University, said, "There is a general interest among biologists in mapping the envelope of life, the outer edges of physical conditions at which animals and plants can exist, and to come to understand the physiological devices by which they are able to do it."

The desert ant, he said, is particularly important because it "occupies the actual edge of the envelope of life and therefore the world as a whole by pushing the limits of tolerance to heat and by collecting heat-killed corpses for a living."

"We are only just beginning to understand how the envelope has, so to speak, been pushed," he said. The lives of these extremists, in addition to teaching biologists how far animals can really go, have provided clues for practical solutions to physical challenges of interest to humans.

Dr. Thomas Eisner, an insect specialist at Cornell University, is also impressed by the special abilities of the desert ants. "The fact that these ants can make it where very few other things can really make it, the very fact that its society can inhabit an area on earth where we humans basically can only exist marginally, tells you it's clearly an animal with incredible adaptive potential," he said.

New work on the behavior of this heat-loving ant published in the journal *Nature* reads like "Ripley's Believe It or Not." The chief author is Dr. Rudiger Wehner, a zoology professor at the University of Zurich in Switzerland, who has made the ants his life's work.

"They're quite amazing creatures," he said of the species, known to ant experts as *Cataglyphis*. "You can say that they are suicidal, as many ants die while they're going out to forage very close to the upper lethal limit of temperature. It's kind of kamikaze. Many never return." Member species of the *Cataglyphis* genus are found from the dry plains of Spain to the Gobi Desert of Central Asia.

If an insect could feel awe, then a Saharan desert ant would tremble at the miles of hot, blowing sand before it, a panorama revealing little shelter and even less life. Yet for its colony to survive the ant must venture out and scavenge from this wasteland at least enough food to raise an ant to replace itself.

Undaunted by their task, these desert ants literally rise to the occasion, hauling themselves up out of their burrows on limbs nearly a quarter of inch long, a great length for an ant. Just by raising themselves a quarter of an inch above the ground, the ants can cool their bodies by nearly 30 degrees Fahrenheit, since even this small distance affords a measure of protection from the surface's intense heat.

The silver ant, which of all the desert ant species endures the highest temperatures, must periodically find and climb small stalks on which to cool off before continuing its hunt. On the top of a flower stalk, the ants can often be seen stretching their front legs skyward to reach the cooler air. The Swiss biologists have found that the cooling-off spells on these refuges are vital if the ant is to survive the heat on its long hunting expeditions.

The silver ant's search for cooler vantage points can complicate the biologist's task, since the ants rush to climb anyone who comes close to them. "We have to be quite careful of the ants when we do our measurements," Dr. Wehner said. "They will run toward you and climb up onto you as the tallest thing around."

In its movements across the burning desert floor, the silver ant tries to touch the ground as little as possible. Like a beachgoer who has forgotten her flip-flops, the ant alternately sprints and hips and hops, sometimes running with two of its six legs held up in the air as it navigates the too-hot sand.

"These ants simply do not walk," Dr. Wehner said. "They sprint." The ants can run a distance that corresponds to 100 of their body lengths per second, an ability that no other animal is known to match. During these spectacular sprints, the ants even hold their breath to conserve every drop of precious body moisture.

The silver ants are social insects and even burst out of their hole all together, but each then goes off alone to find food. The ants scavenge dead and dying creatures on the desert floor, only occasionally taking a healthy prey like a helpless caterpillar or two. Since carcasses tend to be scattered widely and unpredictably, the ants have no need to lay chemical scent trails for their fellows to follow, as most of other ant species do. Instead, they hunt alone, feverishly crisscrossing the sands until they find a prey and can head for home.

Since the ant's travels across featureless terrain may take it up to a third of a mile, the insect might seem at high risk of getting lost. But biologists have shown these ants to be remarkable navigators, guided by the patterns of polarized light that arc across the desert sky.

"We saw these ants winding along until they found something and then, instead of retracing their outbound path, they take a beeline, or an antline if you will, directly back to the nest," Dr. Wehner said. Their method of navigation, known as dead reckoning, requires keeping a constant fix on where one is relative to home.

Polarization of sunlight creates a characteristic pattern in the sky that provides a ready compass for those who can read it. Dr. Wehner and his associates have found that the ants in their foraging make many twists and turns that let them scan the width of the horizon and figure out their position.

Most species make enough turns in their foraging to see the whole sky now and again, but the silver ant runs in straight lines like a missile headed for a target. These straight lines, however, are interrupted by what Dr. Wehner describes as "graceful, little minuets," which are quick rotations to see the map hanging far above.

During the turns, the ants scan the sky with the uppermost part of their eye, a region specialized to read the celestial map. Built as what Dr. Wehner calls a matched filter, the ant's eye is made like a rough map of the sky's polarized light patterns. The ant scans the sky until the map built into its eye aligns with the scene above so as to show maximum brightness.

According to Dr. Wehner, humans cannot see polarized light but can approximate what the ant sees by rotating polarized sunglasses upside down and right side up again. The sky will appear bright at first, then darken as the glasses go upside down, and brighten again as the glasses complete their rotation.

In order to navigate by dead reckoning, even with a celestial compass, the desert ants must continually work out their direction and distance from home.

Far from doing advanced trigonometry in their heads, the ants, Dr. Wehner and his team believe, are practiced in a few simple approximations that serve excellently in getting them regularly back home, or at least very close. Having found the tricks used by these tiny-brained ants to solve complex problems, Dr. Wehner has been able to give clues to computer scientists about how simple machines can undertake difficult computational tasks.

Through what other researchers have called "the patience of Job" and "an intense devotion," Dr. Wehner and his colleagues have precisely mapped the structure of the desert ant's eye, its internal celestial map and how the ant uses its collected data to navigate. Dr. Timothy Goldsmith, Andrew W.

Mellon Professor of Biology at Yale University, said, "Wehner's work is an extraordinarily elegant bit of analysis."

Out in the salt pan desert where the ground is packed hard, researchers spray painted a giant grid along the desert floor. With the grid as a reference, they tracked the ants with the help of a lawnmower-like device outfitted with a variety of lenses and filters that can be positioned over an unsuspecting ant on the ground.

The scientists, who might well look comic to an uninformed observer, run about trying to keep the ants directly under the lawnmower and its filter as they race along their unbearably hot foraging paths. By restricting the ant's view of the sky, the scientists laboriously pieced together the workings of the small navigator's mind.

Most ants succeed in their foraging expeditions, arduous though they are, and return with 15 to 20 times their weight in food over the course of their short lives, which last on average only six days.

After a hard day's hunt, an exhausted silver ant scoots back into its hole, pulling in grains of sand behind it, to close the door for the day on its home. As the evening wind sweeps across the desert and all traces of the nest hole disappear, the ant retires below until the next day, when the sun scorches the world once again to the perfect temperature.

—CAROL KAESUK YOON, June 1992

Honeybees Make Fight
Too Hot for Giant Hornets

THE JAPANESE HONEYBEE has devised an unusual defense against mass attacks by a major predator, a giant hornet. Responding quickly and in large numbers, the bees engulf hornet scouts and generate enough heat to roast them to death, scientists say.

Researchers at Tamagawa University in Tokyo say the honeybees' thermal defense appears to have evolved to counter swarming behavior by the hornets that often ends in the destruction of entire beehives.

If three or more giant hornets, formally called *Vespa mandarinia japonica,* gather at a beehive, they switch from individual hunting behavior to an attack en masse that results in the slaughter of thousands of bees. However, the researchers say, the bees have devised a way of detecting impending mass attacks and blunting them by quickly killing the first hornets to arrive. When successful, the strategy prevents enough hornets from congregating at a beehive to stage a mass attack, they said.

Reporting in the science journal *Nature,* Dr. Masato Ono and his colleagues in the university's Laboratory of Entomology said early observations of the defensive behavior of the Japanese honeybee, *Apis cerana japonica,* suggested that the bees ganged up on attacking hornets and stung them to death. But studies done by the Tamagawa University researchers over the last decade found no evidence of stinging.

Instead, they say, more than 500 bees engulf each hornet in a ball and raise their body temperatures to levels so high that the hornet dies in about 20 minutes. Some defenders die in the struggle against an enemy that is four times the length and 20 times larger than an individual bee. But their bodies are pushed out of the ball and they are replaced by others as the bees turn up the heat of their attack.

With the hornet entrapped, the bees vibrate and quickly raise the temperature of the ball to 116 degrees Fahrenheit, above the laboratory-measured lethal temperature range for the hornet of 111 to 114.8 degrees, the report said. Tests show that the bees, which display unagitated body temperatures of 95 degrees or less, can survive heat of up to 122 degrees.

Earlier studies by the same researchers showed that Japanese honeybees use the same heat-to-kill method against a smaller hornet, although it involved fewer bees to do the job and did not appear part of as intricate a defense strategy as that against the giant hornets.

Dr. Gene E. Robinson, an entomologist at the University of Illinois at Urbana-Champaign who specializes in bees, said the Japanese honeybee behavior was extraordinary and deserved more study. "Group defense among insects is common, but thermal group defense is unique," he said. Dr. Robinson said researchers should look to see if elevations in body temperature are associated with other defensive behavior in insects, which could be the origin of a response that eventually led to using heat in a coordinated attack.

Defense against mass attack by the giant hornets is mainly observed in the autumn, when nests of hornets produce hundreds of new queens and males that require large amounts of protein. Researchers say this food pressure may force the hornets into intensive foraging leading to the gang attacks.

In such attacks, one hornet can kill up to 40 bees a minute with its mandibles, and a colony of 30,000 bees can be killed in three hours by a group of 20 to 30 hornets, they said. After a successful mass attack, the hornets occupy the beehive for 10 days or more, carrying off bee larvae and pupae to their nest to feed their young.

The attack on a beehive begins with one hornet finding it, killing a few bees and taking them home to its young. After several return trips that result in successfully killing a few bees each time, the hornet smears a messenger chemical called a pheromone on the hive, which signals other hornets to join in an attack, the report said. Soon after the marking, nestmates of the first hornet that are flying in the area congregate at the tagged hive and start individual hunting.

This is when the bees must act quickly and decisively, or risk losing their hive in a mass attack, the researchers said. Fortunately for the Japanese honeybees, they also detect the hornets' pheromone and respond by moving more than 1,000 workers to the inside of the hive entrance. More than 100 workers crawl around the entrance and thermal scans of the area show that their body heat has risen above normal. When a hornet approaches, the bees at the opening flee into the nest, luring their enemy inside where it is jumped by a crowd of waiting defenders.

The pheromone-regulated mass attack of the hornets probably evolved to fulfill the need for large amounts of food during breeding season, the scientists said, and as a counteradaptation to the bees' defensive behavior.

—WARREN E. LEARY, October 1995

A Long-Secret Weapon
of Millipede Is Unveiled

MILLIPEDES AND CENTIPEDES are the many-segmented and multilegged animals that live in rotting logs, but there is a huge difference in their strategies for survival. Centipedes are fast-moving carnivores with venom-filled jaws for killing insects, earthworms and other small prey, while the slower millipedes feed on decaying plant matter and have evolved passive defenses against insects, their chief enemies. Most millipedes, when alarmed, curl up like watch springs with their soft undersides hidden and their heads at the center of the spiral. If attacked, they emit toxic compounds like cyanide from glands on each body segment.

But a minute millipede that researchers have found under the loose bark of Florida slash pines appears to have evolved a remarkably effective means of thwarting predatory ants that is mechanical rather than chemical: It turns tail, splaying caudal tufts of detachable hooked bristles that incapacitate its antagonists.

While some tropical millipedes reach a length of nearly one foot, this species, *Polyxenus fasciculatus,* is about an eighth of an inch long. The millipede's strategy was revealed when biologists placed it face to face on a petri dish with tiny ants they had encountered while probing its tree-bark habitat. The scientists watched through a microscope as the millipede swung around and flexed its rear the instant it was touched by an ant, swiping it with bristles that detached and locked onto antennae, mouthparts and legs.

"The ant's first reaction to this insult is to preen," said Dr. Thomas Eisner, a biology professor at Cornell University in Ithaca, New York. "But the shed bristles are a cross between the quills of a porcupine and the hooks and loops of Velcro. The more the ant preens and struggles, the more entangled it becomes." Ants stuck with many bristles will die.

The millipede has enough bristles to repel attacks by several ants, and the supply is renewed when it molts.

The tactics of this millipede, collected by Mark Dreyrup, a staff scientist at the Archbold Biological Station near Lake Placid, New York, are detailed in a paper published in *The Proceedings of the National Academy of Sciences,* one of more than 140 papers on the defense mechanisms of arthropods by Dr. Eisner and his colleagues at the Cornell Institute for Research in Chemical Ecology. Scanning electron microscope images made by Dr. Eisner's wife, Maria Eisner, show the movements of the bristles, which are barely visible at low magnification with standard microscopes.

Dr. Eisner explained that each bristle was tipped with a three-pronged "grappling hook" that anchors on the body hairs, or setae, of the ant, while the shafts were covered with barbs that link the bristles into loose tangles that can cripple the predator. He said virtually any enemy of the millipede, including centipedes, spiders, pseudoscorpions and beetles, would be stopped by the bristles because body hairs are a standard feature on arthropods.

European scientists suggested more than a century ago that the little-known millipedes of the polyxenid group used their caudal tufts for defense in place of chemical weapons, Dr. Eisner said, but the mechanics had never been explained.

He noted, however, that the detachable bristles were not an infallible defense. A Brazilian ant, he said, has evolved specialized jaws that enable it to grab a millipede, kill it with a sting and strip the bristles before eating the animal.

"That's life in the natural world," he said. "For every strategy, no matter how perfect, there is a counterstrategy."

—LES LINE, February 1997

Laboring Ants Can Raise
Brood of Warriors

A NEW STUDY has shown that ants can perform a biological feat long speculated about but never proved: manipulating the expression of genes among developing juveniles so they can increase the ratio within a colony of large, aggressive "soldiers" to more docile workers in order to respond to a threat.

The effect is roughly as if a human colony threatened by pirates suddenly gave birth to a race of towering, muscular Goliaths. In the study, done by a group of European scientists, larvae that would ordinarily have become small "minor workers" instead developed into much larger soldiers. The change was set off by the presence of potential invaders of the same species. And adult worker ants seem to have put the genetic maneuver into effect by changing the food supply of developing larvae to alter the hormonal balance.

The study, published in the journal *Nature,* generated immediate debate among biologists about how the finding fits with an important theory about evolution among ants and other social insects, which suggests that thousands of generations of natural selection fix the ratios of various types, or castes, of individuals. Previous studies of other ant species had failed to find that colonies could alter those ratios.

The study was conducted by Luc Passera, Eric Roncin and Bernard Kaufmann of the University of Paul-Sabatier in France, and Laurent Keller of Bern University in Switzerland. The researchers removed all existing soldier ants from colonies of the species, *Pheidole pallidula,* and then exposed test colonies to each other on opposite sides of a fine wire mesh. The barrier prevented attacks, but allowed the ants to insert their antennae through the mesh to sample identifying pheromones on each other's exoskeletons.

By the fifth week the test groups were producing more than twice as many soldiers as the control colonies.

The mechanics of just how ants alter the development of their young to produce different body types is poorly understood, Dr. Keller said, "but we know there's a difference in hormonal status between juveniles of the two types, and it's very likely triggered by switching to a particular type of food." The hormonal change would in turn suppress or enhance the expression of existing genes, resulting in, for instance, the much larger head of soldier ants.

Deborah M. Gordon, a biologist at Stanford University, compared the phenomenon to an individual animal's being able to alter a body part in response to a threat. Writing a commentary on the study in the same issue of *Nature,* Dr. Gordon pointed out, as an example, that some species of bryozoans, which are primitive coral-like marine animals, will grow spikes in the presence of predators.

"It's a really exciting example of an organism's ability to respond flexibly to its environment," Dr. Gordon said. "But it's especially intriguing because it's the colony that's responding, rather than an individual, by changing the kinds of individuals it produces."

But while Dr. Keller suggested that many other species of ants might demonstrate this sort of plasticity, Dr. Gordon said in her commentary that she doubted that many other species would show the same ability. She referred to a widely accepted evolutionary theory outlined in 1978 by the biologists Edward O. Wilson and George Oster. The theory suggests that such shifts in caste distribution should normally occur only "over many generations" through natural selection. *Pheidole pallidula,* she said, was probably an "exceptional species." She also said soldier ants were costly to a colony because they were larger than worker ants and demanded more food. Furthermore, it may take weeks for a colony to change its population structure. Both facts would argue against quick shifts in caste distribution being common in the ant world.

In an interview, however, Dr. Wilson, who is Pellegrino University Professor at Harvard and an expert on the *Pheidole* genus, suggested that the discovery could fit within the framework of his theory. "The result is not only very interesting, it's entirely reasonable," he said. He noted that his own efforts in the 1980s to induce a similar change in another species of the same genus failed to show a shift in ratios. "This certainly leaves open the question of why this occurs in one species, but not another," he said.

Yet he said he was inclined to think that further research would probably uncover many more species that could alter their colony compositions.

"Each species I have examined has a distinctive caste ratio ranging from about 1 percent soldiers to as many as 30 percent," Dr. Wilson said. And this does appear to be correlated with their response to the environment. For instance, he said, some species live underground much of the time and only use a few soldiers as a sort of home guard. "Others," he said, "forage widely and take along a high ratio of soldiers like an expeditionary force to protect the food they find. But the fact that these percentages are distinctive doesn't exclude the possibility for the kind of flexibility described in this study."

—JON R. LUOMA, February 1996

Honeybees Conceal Reserve Army Among Workers

WHEN HONEYBEES COME UNDER ATTACK, they can repel invaders by calling upon a swarm of reserve forces, soldier bees, that no one suspected even existed, scientists say.

Bee experts have long assumed that foragers, older bees that fly far from the hive to find nectar and pollen, were the chief defenders. Alerted by younger guard bees stationed at entrances and exits, the foragers were believed to drop what they were doing and rally to the defense of the hive when the alarm rang out.

But researchers have new evidence that ordinary honeybees, descendants of insects brought from Europe more than three centuries ago, may have a soldier caste whose main job is to fight, similar to that found among ants, termites and a few other communal insects.

"The defensive response of a colony is more structured than we thought," said Dr. Gene E. Robinson, an entomologist at the University of Illinois who was involved with experiments that unmasked the secret soldiers. "This clearly shows there is a division of labor for defense."

Scientists do not yet know whether the findings have special relevance to Africanized bees, the aggressive honeybee cousins popularly known as "killer bees." These bees, who always seem to be spoiling for a fight, have recently completed a 30-year journey from Brazil and are now laying siege to the southern border of Texas.

"Perhaps colonies of Africanized bees contain a higher proportion of soldiers, or they are better at recruiting other bees as soldiers," Dr. Robinson said. It also is possible Africanized bees do not have this type of social organization at all, he said, and that all or most workers can respond defensively to a disturbance instead of leaving the job to a special group.

The findings by Dr. Robinson and his colleagues, Dr. Michael D. Breed of the University of Colorado and Dr. Robert E. Page, Jr., of the University of California at Davis, could prove highly significant in studying bellicose bee behavior, other experts said.

"This is a very interesting phenomenon," said Dr. Thomas E. Rinderer of the Agriculture Department's Agricultural Research Service in Baton Rouge, Louisiana. "I'm quite eager to see these investigators study African-ized bees to determine whether or not the behavior of soldiers is a chief dif-ference between them and European bees."

Dr. Page has begun preliminary work in Mexico to repeat the experi-ments with hives that have various proportions of Africanized bees.

According to a report published in the journal *Behavioral Ecology and Sociobiology,* the researchers also found that the soldiers were genet-ically different from the guards and foragers previously credited with hive defense. This finding is additional evidence that genetic variability within a honeybee colony plays a role in the jobs individual bees take on, they said.

Less than a decade ago, many experts considered bee colonies to be large, genetically homogeneous units. But recent work by a number of research groups has shown that honeybee colonies consist of subpopula-tions with slight genetic differences that may help determine what special-ized tasks the insects take on.

Although the variables are not well understood, many bee scientists believe that career choice—whether to be a nurse, nectar collector, under-taker or whatever—may be influenced by genetics as well as environmental factors, such as age, in what part of the hive they live and what they eat early in life. Bees, who live from 30 to 48 days as adults, generally change spe-cialties as they age and many end up as foragers.

The hyper-romantic nature of new queen bees is responsible for the genetic variation in bee colonies. In the first weeks of her adult life, the young queen typically mates with a few to a dozen and a half drones. For a year or more, she lays eggs at random intervals and each batch appears to have been fertilized by only one consort, producing lines of workers that each carry characteristics of a single father along with those of the common mother, said Dr. Mark L. Winston, a bee expert at Simon Fraser University in British Columbia.

"The result of having a promiscuous queen is that each colony has numerous, genetically based subpopulations," said Dr. Winston, the author of *The Biology of the Honeybee*. "Each patriline, or the lineage from each drone, appears to have an increased probability to do a task, such as being more likely to forage for pollen, serve in a nursery or, possibly, be a soldier."

Dr. Winston said that while he believes a colony with varied hereditary lines would be better than one that was genetically homogeneous, the superiority of one concept over the other has not been proven.

In the latest work, Dr. Robinson and his colleagues conducted tests that showed soldier bees were genetically distinct from foragers and guards, but they said it was still unclear how this uniqueness influences the insects' actions.

The researchers tested honeybee defenses by agitating hives to draw out defenders. They repeatedly dropped bricks on top of test hives to get the bees' attention, and irritated defenders streamed out to attack leather target patches dangling in front of entrances. To the scientists' surprise, the aggressive fighters proved not to be guards or foragers, but another group of older bees.

The young guard bees, who are about 15 days old, may alert the hive to danger through dancing or releasing chemical signals called pheromones, but they were not among the major stingers in the test. And neither were foragers, old warriors who often display worn, tattered wings from long-distance flying and climbing in and out of nectar sources.

The main defenders proved to be older bees from the colony who were in better shape physically than the workworn foragers.

"If you disturb a honeybee colony in North America, you may get a few thousand bees flying around," said Dr. Breed, the Colorado researcher. "It's very intimidating, but few generally sting. The proportion of bees willing to sting varies from colony to colony and it would be worth studying if there are different proportions of soldiers present."

The identification of soldiers as a distinct group of workers, the researchers said, also may help solve a mystery of bee labor that has mystified observers for decades: the "lazy bee" phenomenon. Many bees spend large portions of their lives in apparent inactivity, they note.

"People who watch observation hives have long been puzzled that the expression 'busy as a bee' doesn't apply," Dr. Breed said. However, if many bees are primed for soldiering, serving as a defensive reserve that can be immediately mobilized, he said, they would give the appearance of inactivity when the hive was undisturbed.

There is also a possibility that bees who choose a military career might serve other functions in a colony, Dr. Breed said, perhaps like human military reservists who can be called up to provide disaster relief.

"A soldier class is not a bad concept, but these bees might have other jobs and spend most of their time waiting, ready to be told what to do," said Dr. Roger Morse of Cornell University, who praised the work behind the new finding.

"You can have a society where everyone has a job to do and does it, or another kind of society where some individuals are flexible and ready to go at a job when needed," he said. "If one approach has advantages over the other, we might find out by studying bees. If you want to understand sociology in this world, there is nothing like the honeybee."

—WARREN E. LEARY, October 1995

Fly Has Ear for Cricket's Song

IT IS BUILT like a cricket's ear. It even hears like a cricket's ear. But this ear belongs to a fly.

Researchers have reported that a fly whose young live off the flesh of male crickets has evolved a unique ear that can hear the trilling summer love songs of its prey. The findings of the study appear in the journal *Science*.

Dr. Daniel Robert, a postdoctoral fellow at Cornell University who was the lead author of the study, said that under ideal conditions the *Ormia* fly's ear, one of the most sensitive ever discovered among insects, can hear the singing of a male cricket from a distance of more than 80 yards.

"It's fairly mind-boggling," said Dr. Ron Hoy, a professor of neurobiology and behavior at Cornell who took part in the study. "It's more sensitive to the songs than a female cricket. It's been able to do the cricket one better." While the antennae of flies can detect the low buzzing sounds made by other nearby flies, they are not built to detect the higher frequency songs that male crickets broadcast from afar to the eardrum-like hearing organs of female crickets. To hear the songs of the males they seek, *Ormia* flies have evolved a kind of eardrum in their chests, the first such ear ever discovered in the fly world.

"I think it's amazing that there's an ear so different on a fly," said Dr. Thomas Walker, professor of entomology at the University of Florida who is a leader in the field of insect acoustics. Dr. Walker said that until this study, it was unknown how these flies, which are used in cricket pest control, managed to hear their prey.

When a female *Ormia*'s ear hears a singing cricket, she drops down onto him from out of the sky, leaving behind a squirming maggot that bores into him and eats the cricket, eventually killing him.

Dr. Robert said the females' zeal in seeking out singers was quite understandable. If not deposited on a cricket, it appears that these voracious maggots soon begin devouring their mother from the inside out.

—CAROL KAESUK YOON, November 1992

3

RITUALS OF
INSECT
COURTSHIP

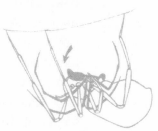

The outcome of courting may be the same in every species, but the preparatory rituals vary widely, and nowhere more so than in the world of insects.

For a female hoping to give her progeny a favorable start in life, a lot of calculations go into the choice of mate. In many species the courtship rituals allow the female to assess the qualities of her suitors and choose accordingly. The males can often improve their chances by acquiring the right nuptial gifts, whether in the form of food or more esoteric items of value.

For fire-colored beetles, the only courtship gift that counts is Spanish fly, a poison known to chemists as cantharidin. Preparations of Spanish fly, prepared from blister beetles that secrete the poison, have long been sold as a purported aphrodisiac, but humans don't know how to use the stuff. The fire-colored beetles do.

The males gather cantharidin from blister beetles and store the poison in special pockets in their heads. The females explore the pockets and only accept males with a good supply of the substance. The females dope their eggs with the poison as a defense against predators.

In the *Gluphisia* moth, the reward package is another scarce resource, salt, which the males gather at great trouble from puddles of water. Another moth, *Utetheisa ornatrix,* has developed a cold calculus for its courtship practice. The female mates with as many males as possible, then selects the sperm of the one she prefers to fertilize her eggs. The trick is made possible by a system inside the female's body for guiding stored sperm packages to her ovary.

Male gifts of food is a courtship practice which the redback spider of Australia has taken to serious extremes: The male himself is the morsel consumed. It is in the male's interest to be eaten, if his goal is to get his genes represented in the next generation:

In unions that do not end in cannibalism the male is rarely the father of his consort's progeny.

These and other forms of courtship, as described in the following articles, show how inventive are insects' variations on the girl-meets-boy story.

———————————

Suicide and Survival: A Spider's Strategy
The male redback spider slowly somersaults into a position that makes it easier for the much larger female to consume him. The gesture prolongs intercourse making it more likely that his sperm will fertilize eggs. The meal seems to dampen the female's ardor making her less likely to mate with another male. It may also be a nutritional supplement helping ensure a big batch of eggs.

Michael Rothman

For an Australian Spider, Love Really Is to Die For

THE VERDICT IS FINALLY IN: Some males will die for sex. They will prance and hop and tap dance and somersault, all for the chance to be devoured alive.

In work that makes grown men cringe and more than a few women chortle, a young scientist from the University of Toronto in Mississauga, Ontario, has demonstrated that it pays for a male redback spider to sacrifice himself to his mate during copulation. Sexual suicide, it turns out, helps to ensure his paternity and discourage the female from mating with his rivals.

The new work, which appears in the journal *Science,* is the first to demonstrate that males can benefit genetically by inviting their mates to consume them. In studying the poisonous redback spider of Australia, whose scientific name is *Latrodectus hasselti,* the researcher, Maydianne C. B. Andrade, has found that those males eaten during copulation sire proportionately more offspring than do the partners that the female spiders choose not to chew.

Scientists have long known that some female insects and spiders will eat the males during or after copulation, the most famous examples being the praying mantis and the redback's American cousin, the black widow spider.

But many researchers have argued that the males are eaten not by choice but because they failed to escape in time from the female's clutches. They have asserted that the males and females have very divergent interests, the males trying to survive to breed again, the female hoping for dinner.

Ms. Andrade, who is now working on her doctorate at Cornell University in Ithaca, New York, builds on the work of Dr. Lynn M. Forster of New Zealand to show that the male redback actively seeks his own doom, slowly somersaulting during copulation into a position that makes it easier for the female to consume him. Much of the time, the female will oblige by slowly liquefying and devouring him as their intercourse proceeds.

"I find the idea of male sacrifice interesting," Ms. Andrade said. "It indicates that cannibalism doesn't always have to be a conflict of interest, of one individual being overwhelmed by another." For the redbacks, she added, "it's obviously a collusion."

The driving force in male redback behavior is, as in all species, competition from other males. A male that arrives at a female's web, eager to mate and be done with it all, may find up to six other suitors milling about her parlor. Or he may strike it lucky and find himself alone.

On reaching the web, he begins his elaborate courtship dance. He bounces up and down on the springy strands. He taps the threads with a foot. He takes a bit of the web and wraps it into a ball, which helps mask the scent of the female's pheromones embedded in the strands that might attract other males to the site. He gets on her abdomen and caresses it repeatedly. The courtship may last three hours or more. Eventually, the female permitting, the male will begin copulation.

Sometimes, though, for reasons that remain mysterious, the female will not consume her lover, no matter how vigorously he offers up his body to her jaws. It is the uneaten male that fathers few if any of her offspring, despite their copulation.

And because the male's life span is just a matter of weeks, and his chances are so slender of making it to another female's web without being killed by predators en route, he has almost no chance of copulating a second time. Paradoxically, then, it is the unlucky fellow who does not die for love.

"It's bad news for the male if he doesn't get cannibalized," Ms. Andrade said.

The suicidal behavior serves several functions, she said. By persuading the female to dine on him, the male prolongs the act of intercourse by several minutes, and the length of copulation correlates to the likelihood of the male's sperm fertilizing the eggs. The act also somehow inhibits the female's ardor. Ms. Andrade has observed that a cannibalizing female is less likely to seek another partner than is the spider that mates without the meal.

Finally, the male's sacrifice may offer some nutritional value to the female, helping to ensure that she lays a large batch of eggs. The donation is a spare one, though, given the extreme difference in size between female and male. The average female weighs 256 milligrams, or about one hundredth of an ounce, while the male weighs about 4.4 milligrams, a mere 2 percent of her mass.

Ms. Andrade's work offers an extreme example of the selective pressures shaping male sexual behavior. While the forces of natural selection result in traits that help keep an individual alive—wings to escape a predator, for example—the forces of sexual selection focus on the ability to seduce and reproduce, talents that may come at the cost of personal survival.

For many insects, the male must impress the female with a very expensive "nuptial gift," a bolus of proteins, nutrients and sperm that may constitute a large percentage of his body weight. The redback goes much farther, and gives his entire being over to the female.

"A lot of evolutionary questions are addressed by looking at extreme examples, and how far evolution might push a particular behavior," said Dr. George Uetz, a spider expert and evolutionary biologist at the University of Cincinnati. "That's what makes this particular spider and this particular paper so interesting."

Dr. Uetz also praised the report for relying not only on laboratory studies but also on extensive observations of spiders in the field. "There's been a lot of criticism of lab work," he said, "because when a male is eaten in a glass enclosure you could argue, well, he didn't have the chance to get away."

Dr. Petra Sierwald, an adjunct curator at the Field Museum in Chicago who studies black widow spiders, says that in that species, too, the male's death is always associated with intercourse, rather than his being mistaken for prey that stumbled into the female's web. But the male black widow does not as clearly covet his own destruction, for he does not perform the male redback's flamboyant lovemaking somersault that presents his body to the female's fangs.

Dr. Sierwald suggests that the entire genus of *Latrodectus*, which includes the redbacks, black widows and other widow spiders, offers so many variations in mating behavior that it is an ideal way to study the evolution of different mating systems and strategies, and the circumstances under which male sexual suicide begins to look like the winning ticket to genetic endurance.

"I'm fascinated by the enormous variety in behavior and morphology that you see in widow spiders," she said. "Though for the public, they may be nothing more than a black or brown spot on the ground."

The public is not entirely immune to the lure of the *Latrodectus*. There is something compelling about the way the spiders make soup of traditional sex roles.

The female redback, for example, appears to have many advantages over her masculine counterpart, beyond her spectacularly greater size. She is black with the brilliant red hourglass-shaped stripe on her back that gives the species its common name; he is a rather feeble shade of white, with a few black or red markings on him.

He lives a few days or weeks as an adult; she can survive for two years. He is lucky to mate with one partner in his lifetime; she may consort with up to three in her youth and possibly more later on. He does not eat as an adult; she can eat him and then some.

"When I talk about my work on redbacks, men shuffle their feet around and look uncomfortable," Ms. Andrade said. "The women say, 'All right, that's great.'" But Ms. Andrade says she often feels sorry for the males. "I once watched one get ripped apart by ants and carried off," she said. "I felt so bad for him. All he was trying to do was make his way to a female's web."

Typical for spiders, the male redback's copulatory organs are at the front of his head, encased in two bulbous structures called palps that look like a pair of boxing gloves. Projecting from each palp is a coil-like organ, the embolus.

During mating, the male inserts the embolus into the female's genitals, located just below where her abdomen connects to her upper body. Using the embolus as a pivot, the male flips over so that his back side lands right in front of the female's mouth. In that position, he can continue copulating through the embolus even if she starts to digest him with enzymes she secretes from her mouth.

The complexity of the male's genitals may explain why male spiders, in general, live such brief lives, Dr. Sierwald said. Females molt, or shed their exoskeleton, several times as an adult, each molting allowing a replenishment of the essential sensory hairs covering the body. But to molt and rebuild the palps and emboli would be too much to ask, and male spiders therefore do not molt as adults. Their warranty expires all too quickly.

Ms. Andrade demonstrated the differences in fertilization success between cannibalized and noncannibalized males experimentally, by using males that had been sterilized with radiation. The sterilization technique did not affect their health or mating ability, but they were incapable of siring off-spring.

By pitting normal males against sterile males and seeing how often the females laid fertilized versus sterile eggs, Ms. Andrade was able to gauge the relative success of one male over another. She also measured how long each copulatory bout lasted, and found that intercourse ending in cannibalization lasted about 25 minutes, while a benign mating endured for 11 minutes.

Ms. Andrade calculated that the difference in reproductive success between the eaten and the spared is considerable. A cannibalized male on average fertilizes 235 of the female's eggs, compared with only 115 inseminated by a male that the female refuses to eat.

The surviving males seem to know the gig is up, and they do not bother seeking a more gluttonous female elsewhere. They remain on the web of their first affair, until they die a few days later of sad old spider age.

—NATALIE ANGIER, January 1996

Spanish Fly Works,
at Least for Fire-Colored Beetles

ONCE THOUGHT to have the power to drive lovers into frenzies of passion, Spanish fly, a purported aphrodisiac that is thoroughly ineffective in reality, has long had a seedy reputation as a sexual snake oil. But in a redemption of sorts, scientists have found that Spanish fly can indeed make males absolutely irresistible—if only to females with six legs and wings.

In a study in *The Proceedings of the National Academy of Sciences*, Dr. Thomas Eisner, a chemical ecologist at Cornell University, and his colleagues have shown that during courtship, female fire-colored beetles taste-test a gland in their suitor's forehead. What they are searching for is a bit of Spanish fly, which in its pure form is known as cantharidin, a toxic substance quite rare in nature. If he has it, she consents. If he does not, forget it.

Males that do mate then give females a much larger dose of cantharidin along with their sperm, providing enough of the potent chemical for the female to lace her eggs with it, protecting them from predators.

"It's like a woman assessing the resources of a man by seeing how many credit cards he carries," said Dr. James Carrel, a chemical ecologist at the University of Missouri in Columbia who is one of few researchers, other than the authors, studying how animals use Spanish fly.

Researchers say the study is the latest to show that fathers of the animal kingdom are providing walletsful of resources to their young in many previously undetected ways—and with good reason.

For many females, it turns out, there is nothing sexier than a good dad.

"This is exciting stuff," Dr. Darryl T. Gwynne, an evolutionary biologist at the University of Toronto in Mississauga, said of the study of the beetles, whose scientific name is *Neopyrochroa flabellata*. "It's really novel. The fact that females are tasting the forehead and rejecting those males that don't provide a good meal of Spanish fly, that's fascinating. The phenomenon of males' doing stuff for females is really becoming widespread. It really does have generality."

The team of researchers, which also included Dr. Scott R. Smedley, Maria Eisner, Dr. Braden Roach and Dr. Jerrold Meinwald from Cornell University and Dr. Daniel Young from the University of Wisconsin at Madison, discovered the importance of cantharidin in the life of these beetles through a series of experiments in which they meticulously traced the substance as it was transferred from animal to animal.

Male beetles, they discovered, would ravenously devour any cantharidin offered to them. Afterward researchers found the chemical in the males' reproductive tract as well as in the gland in their heads, a huge, goo-lined slot running the width of the beetle's broad forehead.

"No brains, just glands," said Dr. Eisner of these males whose heads would seem to have little room for anything but their monumental cantharidin-holding pocket.

When males courted females, they immediately presented the gland. Females would heartily dig in, rooting around with their mouthparts in the male's head for a feeding. Further advances by cantharidin-rich males were usually accepted by females. Attempts at mating by males lacking the chemical were nearly always rejected.

Finally, researchers presented eggs of cantharidin-donating and cantharidinless fathers to insect predators. Eggs laden with the toxin donated by their fathers were well shielded, some rejected as soon as an insect made contact with the egg, presumably getting a taste of the noxious chemical. Those lacking cantharidin, fathered by male beetles with none to contribute, were quickly eaten.

But though the chemical from fathers ends up protecting their offspring, the question arises: Are these males using the protective cantharidin to be responsible parents, or is the protection just a side effect of using cantharidin to attract females?

Dr. Gwynne, who notes that the males may also be doing both simultaneously, suggests that fire-colored beetles may be able to provide an answer to this question, which has vexed so many scientists trying to understand what drives males to provide precious resources to females.

If cantharidin is most important for luring mates, he suggested, males should be dishonest in their advertising.

"If this stuff is so rare in nature," Dr. Gwynne said, "why not cheat the females?" While displaying cantharidin in their heads, they should fail to provide it during mating or provide less than they appear to have.

Responsible dad beetles, on the other hand, should show and share what they have got so as to best protect their young.

While the data are preliminary, Dr. Eisner and colleagues report early indications that fire-colored beetles are honest advertisers, as the amount of cantharidin in a male's gland appears to be an accurate reflection of his other holdings, roughly proportional to the amount he ingested.

Though clearly more useful for fire-colored beetles than for humans, cantharidin was once thought to be a pharmaceutical of great promise.

In the 1600s, Spanish fly, a concoction of ground-up beetles, was one of the most widely used drugs, taken for numerous complaints including baldness, goiter, skin rashes, pain and rheumatism. Named for the beetles used, which were then generically called flies, as were many insects, and were thought to be most prevalent in Spain, the mixture eventually became of intense scientific interest as an aphrodisiac, probably because of the curious effect it had on men. After ingesting the chemical, men can suffer prolonged, painful erections.

So great was the interest in this drug, even the Marquis de Sade did a bit of experimenting, according to Dr. Eisner, loading chocolates with Spanish fly and noting the effects on prostitutes who were subsequently taken rather ill.

Cantharidin eventually proved itself to be much more toxic than intoxicating—as little as 60 milligrams can be lethal. Today it is recommended for little more than the treatment of warts.

One of the mysteries that remains in the story of fire-colored beetles is where they, as well as the many other male insects that are drawn to cantharidin baits, get their Spanish fly in the wild, incapable as they are of producing it themselves.

A few insects, including those known as blister beetles, can produce cantharidin. But Dr. Eisner and colleagues argue that since these fire-colored beetles are not known to prey on other adult insects, blister beetles are not a likely source.

Dr. Carrel suggests that rather than eating them, fire-colored beetles may instead be milking blister beetles for their cantharidin. When blister beetles are disturbed, a leg tugged or bitten, they respond by exuding drops of cantharidin-laced blood from their leg joints in a process known as reflex bleeding. Releasing the toxin, they effectively deter their enemies, including humans who get blisters from the blood, hence the name.

Dr. Carrel speculates that male fire-colored beetles harass blister beetles, then simply lap up the dripping cantharidin, flying off fully tanked for the next female they meet.

Wherever they are getting their cantharidin, fire-colored beetles, by gathering cantharidin and then passing it from male to female to eggs, appear to be good evidence for the newly emerging notion of the importance of chemical webs, according to Dr. Carrel. Researchers studying such chemical webs of transmission are finding that, as in food webs, important chemicals often move from one individual, or one species, to another. But unlike food webs, in which the currency is quickly digested substances like proteins and sugars, these webs involve the movement of chemicals like cantharidin, stable, rare and highly sought-after molecules, as they pass from animal to animal, or plant to animal.

Dr. Eisner, who has made an avocation of tracking the scattered historical entries in which cantharidin displays its powers, noted that even humans had occasional accidental encounters with Spanish fly in nature.

In 1861, for example, a French researcher reported finding soldiers in North Africa showing symptoms of cantharidin poisoning, suffering prolonged, painful erections. The cause of the outbreak of excitement? An inadvertent entry into cantharidin's chemical web as French soldiers dined on frogs loaded with the toxin, the frogs having recently eaten large quantities of blister beetles.

—CAROL KAESUK YOON, July 1996

With Katydids, First to Sing Wins Bride

AMONG THE NATURAL WORLD'S most spectacular sound and light shows are synchronized courtships, displays in which groups of males all croak or flash simultaneously to show their romantic intent to females. In Asia, a group of male fireflies can blink on and off together in the branches of a tree like long, perfect strings of Christmas lights. All around the world, male frogs, katydids, crickets and cicadas fill the night with the deafening alignment of their love songs.

So impressively precise is the orchestration of these courting males that biologists have long assumed that they must be cooperating to keep their signals closely matched.

Early in the study of synchronous choruses, some doubted that the males were actually trying to synchronize, suggesting instead that such patterns were figments of the imaginations of a few scientists.

To test whether males were actually synchronizing, a researcher asked naive observers to listen to different groups of singing crickets. Some groups of crickets had been made deaf so they could not respond to each others' songs, while others could hear. Observers described the hearing crickets as singing synchronously and the deaf crickets as singing every which way. The experiment has made synchrony an accepted natural phenomenon.

But in a study published in *Nature*, researchers say they have found that the males, rather than cooperating, are actually competing fiercely.

The new study, of conehead katydids, suggests that the males are trying to sing just ahead of their fellow choristers, not with them. The alignment of their songs is an incidental byproduct of their competition, but the human ear, with its crude sensitivity, perceives the songs as beautifully synchronized.

"It's an important study," said Dr. Michael Ryan, a behavioral biologist at the University of Texas at Austin. "Now we don't need to try to explain why they're cooperating because in fact they're not cooperating. They're competing."

Dr. Stanley Rand, a biologist at the Smithsonian Tropical Research Institute in Panama, said: "I think it's quite an exciting study, and I think it's going to be influential. It's certainly going to affect the way I go around looking at the question of chorusing in frogs."

Researchers had previously concentrated on hypotheses that attempted to explain what males might gain from synchronizing. Among them was the idea that individual males singing simultaneously with others might be more difficult for predators to locate by their songs. Another hypothesis suggested that a synchronous chorus would preserve intact the species-specific song or flashing pattern, keeping females interested instead of turning them away with a discombobulated mess of jammed signals. But no evidence of any advantage to synchronizing was found in the conehead study, said one of the researchers, Dr. Michael D. Greenfield, a behavioral ecologist at the University of Kansas.

In a series of experiments that scientists have described as "elegant" and "beautifully executed," Dr. Greenfield and Igor Roizen, a computer scientist at the University of California at Los Angeles, deciphered the rules of competition that govern the timing of a male's singing.

Male conehead katydids "sing" their song by rubbing their front wings together, producing a series of fairly regularly timed chirps, two to three per second for several minutes at a time. When Dr. Greenfield played recordings of the songs to katydids in the field in Panama, he found to his surprise that males did not synchronize with the recording. Instead, they appeared to reset their internal chirping metronome to start consistently just milliseconds before their recorded competitor's next chirp. Researchers were a bit befuddled by the males' determination to be the chorus leader, until they tested females to see what songs they preferred. In laboratory tests, when katydid females were given the choice of two speakers playing nearly simultaneous chirps, they consistently nuzzled the speaker playing the song with the leading chirp, even if it led the other by as little as a hundredth of a second. While theories abound, the researchers remain in the dark as to why conehead females should prefer the chorus leader.

While a katydid might always be able to beat out a slow, regular recorded competitor to become the lead chirper, actual competitors can be faster and slower and can themselves change their speed. In the wild, katydids adjust to jump in as the leading chirper only as often as they can. Presumably, he who sings first most, mates best.

—Carol Kaesuk Yoon, August 1993

Nature Switches Roles
for the Mating Game

STEREOTYPED THOUGH THE BEHAVIOR MAY SEEM, in most species it is the female animal who is the finicky one, and the males who vie with each other to attract her favors. Now researchers have found that, among katydids and crickets, nature's normal gender roles are sometimes reversed, with the male standing by as two or more female suitors rear up in a pitched battle for his affection.

And rather than eagerly mating with any female that will have him, as male animals usually do, the male insect will spurn females he finds sexually unappealing.

According to current thinking in evolutionary biology, females must be the choosy ones during courtship because they generally invest more time and resources in their offspring than males do. Given that greater investment, scientists say, females have the incentive to select males that either possess hardy, disease-resistant genes, or that will offer them some sort of prenuptial gift, like food or defensive chemicals against predators.

The males, in contrast, have little to lose and thus will mate hungrily with any females they can.

Researchers believe that female choice has resulted in some of the more flamboyant features seen in males, such as the peacock's tail or the bullfrog's midnight sonatas.

But in recent studies of a newly discovered species of katydid in Australia and the Mormon cricket of the Western United States, Dr. Darryl T. Gwynne of the University of Toronto and Dr. Leigh W. Simmons of the University of Liverpool found that, when food is scarce, the males of both species gain the upper hand. Some of the new results appeared in the journal *Nature*.

Under normal conditions, when food is plentiful and a female cricket or katydid can get all the nutrients she needs from plants, she can produce an ample clutch of eggs on her own and then shop around for a partner to fertilize the eggs.

During those times, the surroundings are alive with males striving heroically to attract mates by rubbing their legs together in a reedy courtship song. So competitive is the playing field that any given male will attract only one female at a time, and some attract no females at all.

During mating, the female mounts the male and receives from him a large capsule, known as a spermatophore, which contains both his sperm and a nutritious blob of protein. After intercourse, the female eats the protein and stores the sperm away in an internal sac for use later to fertilize her eggs. But because there is abundant food in the environment, the protein snack is merely a fringe benefit for the female and has little effect on her behavior.

During lean times, however, females grow desperate for the male's supplementary food. And because only the stronger or more industrious males are able to gather enough food to manufacture a protein-rich spermatophore, the number of eligible bachelors drops sharply.

"Only those males who have been able to produce a spermatophore will call for females," said Dr. Simmons. "So the number of males available to females is much lower when food is scarce."

At that point, a singing male will attract two or more females, which then commence fighting violently with one another. "They really go at each other, scrabbling, lunging at each other, kicking with their hind legs," said Dr. Gwynne.

The females will even continue battling after one has managed to mount the male. "The female has a little spear on her rear end for laying eggs, and she may try to push it between a copulating pair," said Dr. Gwynne. "Sometimes, she's successful."

Studying the Mormon crickets, Dr. Gwynne also has discovered that a male is not the passive recipient of a victorious female's lovemaking. The male will aggressively reject females that he seems to consider a waste of his hard-earned spermatophore.

"After a female has mounted the male, he may pull off and scoot away," said Dr. Gwynne. "She may come after him, but he'll reject her again and again."

Dr. Gwynne has found that the males seem to prefer heavier, meatier females. "The mounting appears to allow him to weigh the female," he said. "Those who are repeatedly rejected are much lighter in weight."

Heavier females carry more eggs than do their slimmer counterparts, said Dr. Gwynne, a feature of obvious appeal to a would-be father.

—NATALIE ANGIER, July 1990

For Fire Ant Queens,
Fecundity Gene Brings Death

AS MARIE ANTOINETTE and King Louis XVI learned only too well, there are times when the mark of royalty has its distinct disadvantages.

Scientists studying imported fire ants, those relentless and ruthless little stingers renowned for ruining many a picnic in the Southern United States, have discovered that the worker ants will selectively execute those queens in the colony bearing a telltale genetic trait. Of great interest to researchers, the gene, called PGM-3, is associated with exceptionally high fecundity.

Fire ants, and many other ant species, often live in nests where there are many unrelated queens busily laying eggs. The new work suggests that the sterile workers are very choosy in deciding which queens will be allowed to survive and reproduce, and which will be exterminated the moment they reach sexual maturity.

By killing off queens that possess the genetic skill for vigorous procreation, the workers seem to be exercising a form of democratic tyranny—ensuring that no single female is able to breed a dynasty big enough to dominate the entire nest.

The report, which appears in the journal *Science,* is among the first discoveries of a gene underlying the sexual and social behavior of social insects, a group of complex and almost militaristically well-organized arthropods that includes bees, termites and some species of wasps.

The precise role of the PGM-3 gene in queen sexual development remains unclear, but the researchers who performed the new experiments suspect that it somehow hastens the maturation of a queen's ovaries and thus would allow her to get a head start on procreation over those queens without the trait—if she were given the chance, that is.

"We think it's very likely that this gene is the only difference between successful queens" and those that are destroyed, said Dr. Kenneth G. Ross of the University of Georgia in Athens. "But this simple difference is likely to trigger a cascading series of biochemical events" that shape many elements of the queen's physiology, including her metabolic rate and her reproductive prowess. Dr. Ross performed the experiments with Dr. Laurent Keller of Bern University in Switzerland.

The researchers believe that the PGM-3 gene also influences the queen's output of pheromones, the chemical messages that serve as the insect equivalent of language. Alerted by the perfume of potentially excessive fecundity, the worker ants can move in for the kill.

"As a social insect biologist, I find this very, very exciting," said Dr. Robert E. Page, Jr., professor of entomology at the University of California at Davis. "This is the only case I know of that clearly demonstrates they've got a gene involved in recognition." And an ant's ability to recognize its peers is at the heart of its tightly orchestrated if sometimes brutal society, he added.

Dr. Edward L. Vargo, a fire ant specialist at the University of Texas at Austin, said: "I think this is a fascinating example of a genetic basis to the success of an insect under reproductive competition. Social insects are usually thought of as organisms that function together, but underneath the harmonious behavior there is serious competition to procreate."

The discovery of queen recognition molecules may also help entomologists who are struggling to come up with new tactics to tame the fire ant infestation, which is reaching pestilent proportions in many Southern states and continues to spread northward and westward. The imported fire ant, or *Solenopsis invicta,* arrived on this continent from South America about 50 years ago, and it has since aggressively outcompeted North America's four native fire ant species. The conquering *invicta* will consume almost anything in its path, from plants and other insects to the rubber expansion joints on bridges. When it stings—which it does repeatedly under the slightest provocation—it lives up to its name and savage reputation.

Researchers hope that by understanding the pheromones through which workers recognize queens, they can come up with some form of fire ant birth control.

In this work, the researchers sought to understand why only 5 to 10 percent of all queens born in a typical multiqueen colony make it to motherhood. They also tried to find out what distinguishes a multiqueen nest from nests in which one female does all the breeding. The two setups appear to be mutually exclusive: In habitats where fire ants live in one-queen colonies, the multiqueen form is not to be found, and vice versa, although the environmental factors favoring one arrangement over another remain unknown.

Dr. Ross and Dr. Keller discovered that the PGM-3 gene spelled the difference between the survivors and the doomed in multiqueen colonies, and between mothers in single-queen and many-queen nests. In ant villages where one queen reigns, the PGM-3 gene is extremely common. The gene is also widespread among the sterile workers of either colony type, although because workers are born without reproductive organs, the gene does not do them much good in the parenthood department.

But in the reproductively active queens of a multiqueen society, the gene is completely absent. "It's quite incredible," said Dr. Ross. "Our sample sizes are very large, but you just don't ever see it. This seems like a rather absolute thing."

The scientists then did experiments to see how workers would treat a PGM-3 queen when confronted with her. The results were brutal.

"The workers will pin her down, pull her out spread-eagle, and start snipping away," Dr. Ross said. "Eventually you'll have a legless, antennaless queen writhing on the ground. They just pick her apart into little pieces."

But her death seems to be the price the colony pays for the privilege of having more than one queen. "If you want a lot of queens, each one has to have low fecundity," Dr. Ross said. "Otherwise you might see the eventual domination of one or a few queens, and the whole system could fall apart." Even from the luckless queen's perspective, then, the sacrifice may not be in vain. Ever the patriot, she dies for the good of her colony.

—NATALIE ANGIER, May 1993

Any Male Moth Worth His Salt Must Be Able to Drink Deep

LEAVE IT TO THE MODERN FOOD SCENE to so tarnish the reputation of a nutrient that the essential is deemed immoral. Fat and cholesterol have been sufficiently vilified that people may forget that it is impossible to build a body cell or whip up a sex hormone without them. As for salt, groceries today are like an Ancient Mariner's nightmare, so drenched in added sodium that you could die of thirst, or at least high blood pressure. Who needs salt, except our spoiled taste buds?

The answer, as a type of indefatigable moth makes spectacularly clear, is any creature that wants to stay alive. Researchers from Cornell University in Ithaca, New York, have reported their detailed analysis of male *Gluphisia* moths, which spend many hours of their brief life span parked beside puddles on the ground, drinking in enormous quantities of water and then expelling it from their behinds with the force of a firefighter hosing down a blaze.

Dr. Scott R. Smedley and Dr. Thomas Eisner have determined that the moths perform this bizarre ritual, called puddling, as a way of extracting sodium from the water and storing it within them. The males later use their personal little salt shakers in their courtship ritual, bestowing the prized ion on females. The females in turn donate the sodium to their offspring—in other words, the researchers say, they "salt their eggs."

The report on moth puddling appears in the journal *Science*.

The male *Gluphisia*'s feat of drinking in huge volumes of fluid far outstrips the best fraternity beer guzzler and offers a striking example of a living creature's fundamental need for salt. Sodium, a mineral, is essential to virtually every chemical reaction that takes place in the body. While animals lived in the seas, they were comfortably immersed in the ion, but once they ventured

145

onto land, the quest for salt became a driving force in the evolution of many species. Terrestrial herbivores that live far from the seaside face a particularly difficult challenge, for the plants they feed on often are low in salt.

Scientists had long observed that a number of butterflies and moths will puddle from standing water—sucking it in, squirting it out—and they suspected the insects might be prospecting for salt. The Cornell biologists also knew that the *Gluphisia* moth had a serious need for salt, because its major food source, the quaking aspen, has considerably less salt in it than the average tree. However, this paper is the first to demonstrate that sodium uptake does occur and to describe the male *Gluphisia's* startling adaptations designed for its athletic passing of water.

Working in the fields around Ithaca at night, Dr. Smedley found groups of the small nocturnal moths puddling away in the mud, some of them for three or four hours without rest. He compared the sodium content in the water they drank from with that in the jet streams expelled from their anus, and he found that despite the speed with which the water moved through the moth's body, the sodium was sucked away en route. Back in the laboratory, Dr. Smedley contrasted the sodium concentration in the tissues of puddling moths with that of moths prevented from imbibing.

Through these and other studies, he and Dr. Eisner calculated the moth's capacities. They found that the male can take in and spurt out 20 jets of water a minute, each jet measuring almost a foot in length and a seventh of the moth's body mass in volume. It squirts out the water in a fast, rhythmic pulse, rather like a lawn sprinkler.

"It's just stunning," said Dr. Eisner. "The rhythm of the puddling is so fixed that we could photograph it by singing a song to the beat of the pulses and then press the shutter on one of the beats."

Over 3.4 hours, the most vigorous moth passed 4,325 jets, amounting to 600 times its body mass. To manage an equivalent act, a human would have to drink and expel 45,500 quarts of water at a rate of 3.8 quarts a second.

The moth makes the most of his tippling, raising his internal sodium levels to more than eight times that of a nonpuddling *Gluphisia*. At the same time, the drinker does not retain significant amounts of other elements in the water, like potassium. He takes everything for a grain of salt.

In all his particulars, the male *Gluphisia* looks like a professional puddler. Instead of the long proboscis typical of many moths, *Gluphisia* has a short, curved beak that opens into a wide oral cleft built for fluid intake. Guarding the hole are interdigitating projections that act as a sieve to keep out bits of sludge that might slow down the slurping. Moreover, the male's hindgut is much bigger and more densely packed with absorptive villi than that of the female, the better to sponge up sodium.

But the male is no Scrooge. In other work, the scientists have demonstrated that the moth gives away about half of his sodium, apparently by incorporating the ingredient into his spermatophore, the salubrious package of protein, nutrients and sperm that serves as his so-called nuptial gift to the female. She then passes the sodium along to her eggs, ensuring the larvae a supply of the ion when they hatch and begin feeding on salt-free quaking aspen. The deal appears to be that she makes the eggs, he adds the spice.

Dramatic though the puddler's methods may be, other animals have shown equivalent passion for salt. As Dr. Derek Denton described in his classic text, *The Hunger for Salt,* many big game animals migrate long distances in search of mineral fields, which are natural salt licks. Elephants, for example, dislodge clay with their toenails and hoist it up to their mouths with their trunks to lick the salt within. Monkeys have been seen to dip their potatoes in salt water before eating. Other animals frequent termite mounds for their salt, because the upturned earth in the mound brings salty deposits to the surface.

Hunters hoping to attract deer often put up salt licks. People may inadvertently attract animal visitors to their homes should they urinate outside, say, on a porch post. Porcupines have been known to chew the impregnated posts apart to get at the residues of salt, Dr. Denton writes. In fact, anthropologists suspect that the reindeer, one of the first animals to be domesticated, was initially drawn and bound to human encampments by the ever-replenished supply of human urine.

In contrast to herbivores, carnivores have a slightly easier time getting salt, for blood and other body fluids they consume have sodium in them. Humans, as omnivores, have a long history of honoring salt as one of life's imperatives. Magical powers were attributed to it, while its pungent flavor led to the metaphorical use of the word to indicate zest or wit. Salt also has been associated with fertility and fecundity, and indeed a pregnant woman needs more salt for her developing fetus. The word "salary"—one's very livelihood—comes from the Latin for salt. When the Europeans first started colonizing Africa, they used salt as a bargaining chip. The British set off riots in India earlier in this century when they put a tax on salt. Some authorities have even suggested that whatever cannibalism has taken place in human history may well have been induced by a desperate need for scarce minerals like salt.

The very contemporary plague of high blood pressure among black Americans has been linked to salt as well. Some evolutionary biologists—noting that African blacks do not have the high rates of hypertension that African-Americans have—have suggested that the black slaves who survived the ghastly ocean crossing aboard ships where they had little access to water or food were those with an exceptional ability to store sodium. By this theory, their descendants inherited this predisposition, which is of little use and possible harm in a land awash in salted foods.

—NATALIE ANGIER, December 1995

Moth Mothers Mate Often,
Then Select the Father

IS THERE NO END to the finickiness of the female heart? It would seem quite enough that a female bullfrog requires of her suitors a night of aerobically draining serenades to prove their worthiness as lovers. Or that a female pit viper first will watch impassively as males compete like frenzied, one-armed wrestlers for the privilege of wooing her, and then demand that the victorious male spend long hours rubbing her with his chin and flicking her with his tongue before agreeing to the big event.

Now scientists have discovered strong evidence that a female's discriminating taste in mates can continue even after intercourse is through. Biologists studying an orange and black moth common to the Southeast, Mexico and the Caribbean have found that a female will mate with more than one male and then actively select from the various offerings the sperm of the biggest male she has dallied with—the mate presumably bearing the most robust genes for her offspring.

Studying the *Utetheisa ornatrix* moth, Dr. Thomas Eisner of Cornell University in Ithaca, New York, and his colleague Dr. Craig W. LaMunyon, now of the University of Arizona in Tucson, found that female moths engage in promiscuous sex to gather the desirable packets of defense chemicals and nutrients that accompany the male's sperm during intercourse. But once the various nuptial gifts have been collected, the female uses internal muscles of her reproductive system to push along the sperm of the biggest male toward her eggs, while reabsorbing the semen of a lesser male before it has a shot at fertilization.

The study, appearing in *The Proceedings of the National Academy of Sciences*, offers the first real proof that females engage in postcopulatory sexual selection, picking the sperm they want and rejecting the sperm they do not, and altogether assuming enormous control over their reproductive affairs.

"The more we learn about this moth, the happier my daughters are," Dr. Eisner said. "They tell me this is the ultimate liberation story."

Other evolutionary biologists with a long-standing interest in the role of female choice in shaping animal appearance and behavior praised the new work as a gem of an experiment.

"I think it's just wonderful, a significant piece of work," said Dr. Randy Thornhill of the University of New Mexico in Albuquerque. "They certainly did some nice maneuvers to look at the effect of the female in manipulating the paternity of her offspring." Dr. Thornhill, who was the first to suggest that females might engage in postcopulatory selection of sperm, said the strategy was likely to be widespread in the animal kingdom. Wherever there is something for females to gain from mating with more than one male, he said, there will be incentive for them to devise ways of gaining the advantage of philandering without paying the cost in possibly bearing the feeble offspring of feeble males.

The new work is the latest chapter of an in-depth investigation of mating schemes of the *Utetheisa* moth. In past work, Dr. Eisner and his co-workers showed that the males invest enormous amounts of time and effort gathering defensive alkaloid chemicals from bean plants to pack along with semen and protein into spermatophores, the so-called nuptial gifts the males offer to females as incentives for mating. The female needs the defensive chemicals to spread on her eggs as protection against predators. In addition, she craves the extra nutrition and calories found in the spermatophore as a way to heighten her own strength and fecundity.

To demonstrate that he is endowed with a generous spermatophore, a courting male will extrude from his head brushes scented with a whiff of the defensive alkaloid, and lightly whisk the little brushes against the female. If she agrees he is properly furnished, the female allows the male to mount her and ejaculate the spermatophore, a considerable feat.

"The spermatophore in these moths is huge," said Dr. LaMunyon. "It's 11 percent of the male's body mass, and it's quite a large amount of stuff to ejaculate. It's equivalent to a 180-pound man having a 20-pound ejaculate."

But while the male may hope the donation will pay off in his being represented among future generations, the female seems to have an agenda of her own. The scientists have determined that while a female will mate with up to 13 males during her fertile period if she gets the chance, her eggs end up being entirely or largely inseminated by one male.

Allowing 53 female moths to mate with two males apiece, and then examining the paternity of the offspring through telltale genetic variations, the scientists discovered that in 70 percent of the cases, the bigger male sired 100 percent of the offspring; and in all but two of the cases the heavier male fathered the great majority of the offspring. It did not matter if the bigger male was first or second in the mating lineup, his sperm cells almost inevitably found their way to the eggs.

To explore whether the outcome of the fertilization is the result of the superior mobility of the larger male's sperm, or whether the sperm's progress is under the female's direct control, the biologists allowed females to mate, and then put the moths under anesthesia. The anesthesia was known to be of a type that does not inhibit sperm mobility, but instead acts only on the muscles of the female's genital tract. In theory, if the sperm themselves were competing by swimming more or less rapidly in greater or fewer numbers, then the female's musculature should have little effect on the sperm wars.

Under these anesthetized conditions, the sperm of both males stayed right where they had been deposited, near the portal of entry. Without the female to guide the sperm toward her egg chamber, or to push it aside as unworthy, nothing happened.

Dr. Eisner said they had not ruled out the possibility that the sperm were somehow disoriented in subtle and indetectable ways, but he said, "this is the first case with reasonable data to say the female is making the choice." By this scenario, the female detects the largest male because the largest male also makes the largest spermatophore. The hefty packet in turn stretches the genital canal, a distension that is in a sense memorized by the female. Knowing the relative mass of her paramours, the female can choose which sperm sample to channel toward her eggs by appropriate squeezing of the muscles, diverting unsavory sperm toward side chambers along the way.

The researchers said they did not yet know if the larger males the females clearly prefer actually spawn superior children—for example, moths that are especially good at finding food sources and gathering defensive chemicals from plants. Nor have they figured out what the male's counterstrategies against exploitation may be.

"A lot of these males are being cuckolded for their alkaloid when they don't sire offspring," said Dr. LaMunyon. "The question we still have is, how do they try to fight back?"

—NATALIE ANGIER, May 1993

Biggest Evolutionary Challenge May Be Other Half of the Species

IF THERE WERE EVER to be a truce in the raging battle of the sexes, it ought to be during that magical moment when the elaborate dance between coy females and insistent males has at last been resolved in the communion of copulation.

Instead, a study indicates that it is during the actual act of mating that males and females are having what appears to be their most vicious, knock-down drag-out fight of all, an evolutionary struggle cloaked in the deceptively cooperative act of consummation.

One scientist has managed to uncover the power of this battle in an ingenious study of fruit flies published in the journal *Nature*. He did it by preventing females from evolving, effectively tying their hands in the evolutionary struggle.

With male fruit flies unleashed to do their best, or worst, they quickly evolved a number of nasty tricks, including a seminal fluid so potent that it was not only much more effective at preventing females from producing offspring with any other males, but it actually shortened the females' lives.

The work is the latest and most striking of a growing number of studies on the power of the conflict between males and females of various species during mating. Some of the most interesting and controversial studies are revealing aspects of human mating battles.

The study unmasks not only the intensity of the fight but also the speed with which the sexes can sprint to outmaneuver one another. It also suggests that of all the many trials in life, whether it is the struggle against the elements or the struggle to best one's competitors, the most challenging and never-ending may well be the struggle against one's own mate—a finding that may come as a surprise to no one.

"It's the magnitude that's made people sit up and pay attention," Dr. Andrew Clark, an evolutionary geneticist at Pennsylvania State University, said of the antagonism revealed by the fruit fly study. "This illustrates it in a really clear and stunning way."

Dr. Randy Thornhill, an evolutionary biologist at the University of New Mexico in Albuquerque who studies mating behavior in insects and humans, said of the fruit fly work, "It's very elegant, brilliant experimentation."

Dr. William Rice, an evolutionary geneticist at the University of California at Santa Cruz, started with the notion that males and females coevolve, each developing physiological attack and defense strategies in mating, much the way predators and prey adapt to each other.

To begin to tease apart just who was trying to do what to whom, Dr. Rice allowed males from the adapting strain to mate with females from a different strain, the so-called target strain. In order to keep the target strain from adapting, females were always taken afresh from the strain and they and their resulting daughters discarded. With females sent down this evolutionary black hole, it was as if the target strain never encountered the adapting males and so was unable to make an adaptive response to them.

Dr. Rice kept only sons and used some clever tricks to breed out the mother's genetic contribution to these male offspring, a feat not possible in most organisms. The resulting males were then allowed to face off in a new round with a fresh batch of females from the target strain, always evolutionarily naive to their ways.

While no one can predict exactly what females might have done if allowed to evolve, researchers say the sorts of counterinnovations that females could have come up with include chemical defenses against the seminal fluids, or changes in their receptors for the seminal fluid that would alter their reaction to it.

What Dr. Rice found was that in just 41 generations, the adapting males evolved to be much more successful at producing offspring, at enticing already mated females to remate with them and at keeping females from reproducing with other males than males that had to cope with corresponding, and frustrating, evolutionary change in females.

The males' gains came at a high cost to females, whose lives were shortened both by the increased numbers of matings and by the apparently increased toxicity of the seminal fluid itself. Dr. Rice suspects that males had evolved a change in nature or amount of the fluid, which researchers know contains drug-like proteins with the ability to alter females' behavior to a male's advantage, for example delaying a female's remating, or making females lay more eggs sooner. The females presumably find an advantage in spreading out their eggs, mating often and with many males.

While researchers easily agree that such conflicts are likely to be at play in the internal affairs of many animals, the extent to which these same forces impinge on sexual behavior in humans—a delicate subject under any circumstances—remains a matter of no small controversy.

It is not that males and females, whether humans or fruit flies, must by their nature always be in conflict. As Dr. Rice puts it, when animals are truly monogamous, many conflicts evaporate, as what is good for one is good for the other.

When males compete for groups of females, as among lions or elephant seals, they use outright physical force. When females mate with more than one male, seeking the best genes for their offspring, they create a biochemical arena of competition. The consequence, in theory at least, is chemical warfare, and the weapons are sperm and seminal fluids, in both fruit flies and humans.

Dr. Robin Baker, an evolutionary biologist at the University of Manchester in England, is one of the few researchers doing experiments on the evolution of human mating. He says his work suggests that despite the complex explanations that humans come up with for their sexual behaviors, he sees the same evolutionary conflicts operating in humans as in Dr. Rice's fruit flies, with similar strategies and counterstrategies.

In research described as "new and revolutionary" and that others have yet to try to replicate, Dr. Baker and a colleague, Dr. Mark Bellis of the Public Health Laboratory in Liverpool, England, found evidence that women promote sperm competition just as do females of other species.

Based on surveys of sexual activity in Britain, the researchers found that while women tended to spread matings with their regular partners fairly evenly over their menstrual cycle, they were much more likely to engage in double matings, seeking sexual intercourse with men other than their main partner during their time of highest fertility.

It would seem that in all this, the human idiosyncrasy of contraception, with its spermicides and barriers to the cervix, might add confusion for researchers. Interestingly, Dr. Baker says they found that women who were engaging in such double mating were also much more likely not to use contraception, setting the scene for sperm competition.

Dr. Baker, whose book *Sperm Wars: The Science of Sex,* was published by Basic Books, notes, however, that these are not conscious strategies.

"I wouldn't expect women to rationalize their behavior along the same lines that I'm describing," he said. But, he said, though an individual might attribute her action to any number of reasons, the women's collective actions speak louder than any words.

But are men up to the evolutionary challenge?

Dr. Baker said the evidence is good that not only are men up to it, but they are well armed for sperm competition, since they have been dealing with it for some time.

By counting sperm flowing back out of a woman after sex, researchers found evidence that men's ejaculates had the power to block the retention of sperm from subsequent matings. This power appears to remain intact for as long as eight days. The study indicates that it is something in the sperm or seminal fluid itself, since if the man uses a condom during intercourse, no such blocking effect is seen.

Their interpretation is simple. Men's seminal fluids have evolved, just as have those of many other animals, to protect against their sperm's being jeopardized by future matings by their mates.

In what he acknowledges are their most controversial findings, Dr. Baker says they even have evidence that men produce different types of sperm with different functions. While some are adapted for the expected work of fertilization, they are accompanied by others, dubbed "kamikaze" sperm, which appear to be built to block the passage of or even kill competing sperm.

"Clearly sperm competition is a paradigm that applies to humans," Dr. Thornhill said. "There's no reason to expect humans to be exempt."

Researchers in the new field of evolutionary psychology, meanwhile, have moved well beyond consideration of such simple acts as sexual intercourse to dissect the finer details of human mating behavior. They say they are also finding evidence of the same male-female evolutionary conflicts.

"It's a scientific revolution," Dr. David Buss, an evolutionary psychologist at the University of Michigan in Ann Arbor, said of this recently named science, which sometimes seems an uncomfortable mixture of scientific method and cocktail party conversation.

Describing a behavior that he says has clearly been shaped by evolutionary conflict between the sexes, Dr. Buss said, "Women like having long discussions with their friends where they will recount in great detail the exact conversation they had with a guy and talk about what was his tone of voice, the expression on his face. They'll analyze these in excruciating detail to determine his state of mind. Men almost never do that sort of thing."

His explanation is that because women desire committed relationships in which men will invest important resources, they have evolved to engage in such conversations as sophisticated "assessment mechanisms" for judging the commitment level of a potential sexual partner.

Men in turn, he says, seeking short-term sex for its higher reproductive payoff, have evolved to become better and better at deceiving women into thinking they are more committed than they are. "He looks the woman in the eye ever more intently and declares his undying love," Dr. Buss said. "It is a product of natural selection, all of our psychological mechanisms are."

Dr. Thornhill agreed, saying, "The problem women have and have always had is figuring out if the old boy is going to invest or not. They pry and they pry and they pry and they pry to try to get the guys to come out with how they feel. Men are very reluctant to discuss their feelings and they do that to keep women guessing."

There are many, however, who balk at even the approach of analyzing human mating behavior just like that of any other animal.

"Our behavior is not tied to our biology," insisted Dr. Garland Allen, a historian of genetics at Washington University in St. Louis, saying such studies ignored the powerful effect of culture on human behavior. "People get all hung up because they make these superficial comparisons, comparing animal and human behaviors and think they're talking about the same evolutionary causes."

Some scientists, he said, will describe other animals' behavior in human terms, using words like rape, even divorce and prostitution, to describe the behavior of ducks, for example. "Then these scientists turn around and say these human behaviors are so similar to what's going on in animals, it must be the same."

Still others have objected to research on human sexual behavior on moral and religious grounds, deriding work like Dr. Baker's as nothing more than scientific voyeurism.

Putting such emotionally charged issues aside, researchers do agree that while some things are known—for example, there is no evidence that men's ejaculates are toxic to women—until repeated, controlled experiments are done, any other thoughts about how much alike a fruit fly and a man really are during sex remains simple, if entertaining, speculation.

—CAROL KAESUK YOON, June 1996

4

LIVING
ARRANGEMENTS

Many insects are able to form complicated relationships with others of their own species or with different plants and animals.

Ants, bees, wasps and termites have learned to exploit the advantages of sociality. By creating societies with castes specializing in different lines of work, these insects have produced remarkable accomplishments, such as the fungus gardens of leaf-cutting ants and the soaring mounds of termite architects.

Recently biologists have found that other insects too have caught on to the advantages of sociality, including such obscure species as thrips and the ambrosia beetle.

Insects are also adept at evolving relationships of mutual benefit beyond their own species. Several kinds of butterfly caterpillar have learned how to provide benefits to ants in return for their protection against the caterpillar's many predators. In certain species of the Riodinid family of butterflies, the caterpillars have sound-producing organs that they strum to attract ants. The ants feed on amino acids secreted by the caterpillar and in return they protect it from wasps.

The fruitful alliance between plants and insects is well known, but in a further twist to the story, biologists have found that certain plants even change the color of their flowers as if to tell visitors where the nectar is best.

Insects' lives are filled with savage conflict, but they also exploit the benefits of cooperation.

Social Castes Found to Be Not So Rare in Nature

FOR YEARS, an exclusive club made up of ants, bees, wasps and termites has ruled as the social elite of the animal world. The insects' well-run colonies of sterile workers devoted to serving their fat, fertile queens so impressed biologists that they were given the scientific classification of social or "eusocial," a distinction even humans cannot claim.

But new studies indicate that many more species besides these well-known examples may be living eusocial lives, giving up their own chances of reproduction to help foster the offspring of others.

Scientists have now discovered eusociality in two species of thrips, a tiny insect. The eusocial thrips live as colonies in plant galls, defended by armed soldiers that never reproduce and will fight to the death against invaders. Eusociality has also been observed recently among a species of ambrosia beetles. The beetles live deep in tunnels inside eucalyptus trees, their sole purpose being to build and maintain their queen's labyrinthine world, never reproducing themselves.

Along with tree-dwelling aphids and the bizarre naked mole rat, these species are helping to swell the ranks of those known to be eusocial.

Biologists say that the new additions may provide an important key to the puzzle of how such societies have come to exist. And researchers are happily predicting that many more such surprises are in store.

"A lot more animals are going to turn out to be eusocial," said Dr. Bernie Crespi, assistant professor of evolutionary biology at Simon Fraser University in British Columbia and the discoverer of the eusocial thrips. "People have thought that everything had been more or less discovered, but now that appears not to be the case at all."

Deborah S. Kent, a researcher with the Forestry Commission of New South Wales in Australia, discovered the ambrosia beetle's self-sacrificing habits by accident.

"We weren't even looking for the beetles being eusocial," Ms. Kent said. "We were looking at them because they're an economic problem in eucalyptus plantations and regrowth and have been for the last 40 years. People had already studied them but never recognized what was going on. The more people look the more they'll find."

Dr. William D. Hamilton, professor of evolutionary biology at Oxford University and a pioneer in understanding the evolution of eusociality, has a great respect for the phenomenon.

"It's the highest form of sociality," he said, "and it's what enables the ants to dominate the tropics. No organism but man so dominates everything that goes on in a tropical rain forest." Dr. Hamilton attributes the ants' success to their ability to organize their activities.

Dr. Hamilton was one of the first researchers to begin to solve the paradox of how the evolutionary struggle to survive and reproduce could give rise to creatures that never reproduce and spend their lives altruistically helping others to survive.

Ants, bees and wasps, known collectively as the order Hymenoptera, all share an unusual genetic system in which males carry only half the normal number of chromosomes, receiving a single set from their mothers. But a hymenopteran female, in the more conventional fashion, receives one set of chromosomes each from her mother and father.

Dr. Hamilton was the first to point out that the unusual genetics of hymenopterans have the curious effect of making a female more closely related to her sisters than to even her own children, since she shares three quarters of her genes in common with her siblings, but only half in common with her offspring. As a result, in order to propagate more of her genes into future generations, she can do better by staying in her mother's nest and helping to raise sisters rather than reproducing herself.

The idea was a breakthrough in understanding eusociality, but not a complete answer. While all hymenopterans have the same genetic system of sex determination, not all are eusocial. In addition, termites, naked mole rats and ambrosia beetles all carry the normal ratios of chromosomes from their mothers and fathers. So scientists have continued to search for a factor besides genetics that lies at the heart of the drive to become eusocial.

The new examples of eusociality hint that the nature of an animal's home may hold the answer.

As more eusocial species come to light, researchers note that a common thread is that of living inside a "valuable resource," like a hive or nest, that has taken much labor to build and maintain, and is valuable enough to die for.

In the near-deserts in eastern Australia, acacia trees dot the landscape, some daintily ornamented with small football-like shapes, inside of which hide the world's two known species of eusocial thrips.

Early in the growing season, female thrips, perhaps an eighth of an inch long, arrive at an acacia plant and begin eating one of its leaf-like stems. The stem will slowly develop into a gall in a football-like shape around the insect, creating a food-filled cavern in which she and all her offspring can live, safe from predators. Some of her offspring develop large front legs, armed with daggers pointing inward, which these soldiers use to kill those who threaten their home.

It turns out that there is only a short period during which a thrips can induce a plant to form a gall. Once that time is past, the thrips in the area must fight it out over those galls that have been created.

Dr. Crespi said the thrips were constantly fending off invaders, some of whom are other thrips that cannot make homes of their own and are specialized in stealing galls, as well as predators who come to eat the thrips themselves. Dr. Crespi's research on these creatures, financed by the National Geographic Society, was a decade in the making.

Researchers say the intense pressure to protect the all-important home base has driven the thrips, as well as others, to evolve a eusocial lifestyle, complete with kamikaze soldiers that will defend it.

Similarly, eusocial aphids, like those discovered by Dr. Nancy Moran, an evolutionary biologist at the University of Arizona at Tucson, patrol and defend the tiny galls they have formed on the cottonwood trees of Arizona; the galls are both a home and a source of food.

Dr. Moran said the galls had only a single exit hole, exactly the right size for a soldier aphid to exit for her patrolling duty. When someone or something disturbs a gall, which can contain thousands of aphids, the soldiers come out to attack and have been known to jam their plant-sucking mouthparts into human as well as insect intruders. Other aphids use horns on their heads to deter attackers.

"I was excited to find it," Dr. Moran said of the North American eusocial aphid. "It was sort of a mystery previously as to why there hadn't been any discoveries in North America, nothing besides hymenopterans and termites. There's more out there, for sure. It's mostly a matter of having someone there to look for it. I was on a hike when I found these."

While valuable, the homes defended by aphids and thrips pale in comparison to those fashioned by the ambrosia beetle. These beetles build extensive galleries of tunnels in the inner deadwood of eucalyptus trees.

When a female begins to build her empire, she works alone for more than six months drilling an entrance hole into the tree. It is often more than a year later when she finally lays the eggs that will produce the help she so desperately needs to continue her tunneling work.

Ms. Kent said these enigmatic beetles were named after the food of the gods because biologists could not figure out what the insects were eating. Researchers now know that the beetles feed upon a fungus that they cultivate in their elaborate tunnel system. Ms. Kent describes the fungus as a moist, black sheen on the tunnel walls.

Researchers identify trees containing a beetle colony by tiny entrance ways, a tunnel of eucalyptus resin that sticks out of the tree like a spout. Unable to observe the beetles without killing the trees, researchers collect their samples by chain saw. After surveying more than 300 different colonies, Ms. Kent said she had found that the fungus galleries created by the beetles could be maintained for nearly 40 years. Scientists do not know how long the beetles or their queen live.

The importance of the galleries, not only for food production, but for safety, has also become clearer to scientists. Ms. Kent said that in all her surveys, only one predator was found inside a beetle home.

Naked mole rats, whose lives were described in detail in *The Biology of the Naked Mole Rat* (Princeton University Press, 1991), also build cavernous tunnels in which they live and love. Now a popular zoo animal, the once mysterious creature, nearly as ugly as it is intriguing, is the only eusocial mammal known.

These buck-toothed rodents dig tunnels in the East African savanna, inside of which reign a reproductive queen and several stud kings. While the number of these mole rats can range up to 300 in a single colony, many of these inhabitants will never reproduce.

Many questions remain for researchers about the origins of eusociality. In particular, the naked mole rat has kept biologists asking why no other mammal, including humans, has evolved sterile, worker castes.

—CAROL KAESUK YOON, February 1993

Caterpillars Sing to Survive

In a striking example of ecological complexity in tropical forests, caterpillars of many butterfly species in the family Riodinidae make sounds to attract ants. In return for a supply of amino acids, the ants protect the caterpillar from predatory wasps. Without the ants, a wasp will quickly carve the caterpillar into pieces which it removes to feed its young.

The only North American butterfly symbiotic with ants is the Karner Blue, above, now endangered. Its caterpillar sounds to ants, but the mechanism is unknown.

166

Patricia J. Wynne

To Drive Away Wasps, Caterpillars Recruit a Phalanx of Ants

MAYBE IT WAS INEVITABLE that singing caterpillars would be discovered by someone whose heroes are Charles Darwin and Charlie Parker.

The music made by the butterfly larvae would never pass for Parker-style bebop. But Dr. Philip J. DeVries, former jazz guitarist and present-day lepidopterist extraordinaire, has found that they do make rhythmic calls to rally their bodyguards: ants that protect the caterpillars from wasps that would otherwise kill and literally butcher them.

This relationship among caterpillars, ants and wasps is typical of the elaborate interdependencies that have evolved among species of the tropical forest. The interrelationships are more marked than those of species in temperate zones, for reasons that are still not fully understood. It may well be a combination of the forests' wet warmth, their antiquity and the year-round stability of the tropical climate that has allowed evolution to spin such intricate ecological webs.

Whatever the reason, biologists are accelerating their efforts to explore this irreplaceable treasure-house of evolution in the face of widespread destruction of the forests.

Dr. DeVries, who as an independent scientist was awarded a MacArthur Foundation grant in 1988, has found that caterpillars of many species in a certain family of tropical butterflies, called Riodinidae, are equipped with minute sound-producing organs. Two of the organs, just behind the head, are tiny grooved rods that look much like güiros, Latin American percussion instruments made of gourds. The others, jutting up from the head in serried ranks, look like nothing so much as guitar picks.

While the caterpillar moves its head in and out, the two rods, called vibratory papillae, beat a tattoo on the guitar picks. This, Dr. DeVries found by recording and analyzing the result, produces 23 pulses of sound per second at an average frequency of 1,877 hertz.

"It's like running a comb over the edge of a table," Dr. DeVries said in a recent interview at the American Museum of Natural History in New York, where he is a research associate.

Humans could hear the sounds if they were transmitted very far through the air. But they are not. Instead, they are borne along by the leaves and stems of the plant where the caterpillar has taken up residence, and from there to the ants.

The mechanism comes into play as part of an elaborate symbiotic relationship in which caterpillars, ants and plants are linked in an evolutionary pact of mutual aid and sustenance. The caterpillars feed on newly sprouted leaves of many species of plants and also drink nectar exuded by the leaves. In turn, the caterpillars dispense life-sustaining amino acids from two tiny protuberances near their tails. The ants feed on this.

"They do a little drum paradiddle on the back of the caterpillar and it produces the amino acids," said Dr. DeVries.

In return, the ants protect the caterpillar from wasps that walk up and down the plant, looking for caterpillars to kill and cut up for food. "The ants are very territorial about food resources and also extremely good visually," said Dr. DeVries. "They will run out and bite the wasp on the leg. The wasp doesn't want to bother with it, and goes away. It is like having a German shepherd or a Doberman around to protect your kids."

The caterpillar also secretes a chemical, in little puffs, that puts the ants on instant alert. "The ants snap their mandibles open and freeze," Dr. DeVries said. "At a visual stimulus, they will bite."

The caterpillars' sound-producing organs, he believes, are a means of keeping the protective bargain permanently in force. "I view them as tools that caterpillars have evolved to maintain ant associations at all times," Dr. DeVries said. "It does little good if you're associated only loosely with ants in the face of wasps. It is the combination of factors that keeps ants around them all the time."

These sorts of symbiotic relationships are common in tropical forests, ecologists are coming to believe, because of greater sunlight, higher rainfall and a broadly more stable climate than exist in temperate zones. The combination allows "a fuller expression of evolution and the buildup of extremely complex biological communities," says Dr. Edward O. Wilson of Harvard University, an authority on ants and an advocate of saving the rain forests.

Tropical species, he said, tend to be packed tightly together in highly concentrated habitats that in many instances are extremely circumscribed— a single mountainside, for instance, or the headwaters of a single river. During ice ages, when life in temperate zones is largely wiped out, many tropical species abandon the tropical lowlands, which turn from forest to grassland or savanna, and retreat to wet refuge forests in the mountains. This affords them time and continuity of existence, thus aiding the evolution of intricate interrelationships.

Even in warmer eras, like the present, species in the temperate zones are not as tightly packed together as in the tropics, and there is less opportunity for elaborate symbiotic relationships to develop.

Singing caterpillars are just one illustration of "extreme forms" of symbiosis in the tropics, said Dr. Wilson. "I could recite a volume of examples" like them, he said.

Insects have long been known to communicate—through chirping, buzzing and behavioral signals—for mating purposes and as a warning to predators. Likewise, vibratory papillae had earlier been observed on caterpillars, and the possibility that they might convey vibrations to ants had been suggested. But Dr. DeVries is believed to be the first to show that caterpillars do in fact call to ants, to describe the mechanism and to document the caterpillar calls with tiny microphones and recordings.

His findings are also believed to be the first to show communication between different insect species in a symbiotic relationship. Dr. DeVries published his findings in the journal *Science* and has been expanding them ever since.

The paper in *Science* "stopped me in my tracks," said Dr. Thomas Eisner, a biologist at Cornell University who is an expert on insect symbiosis and communication. "All of these gives and takes between insects are fascinating, and he's added a beautiful component to it."

Since the discovery, which came from studies of a species of Riodinid caterpillar called *Thisbe irenea,* Dr. DeVries has auditioned other Riodinid caterpillars of the neotropics as well as members of another butterfly family, the Lycaenidae, from around the world. About 50 more species were found

to produce the call, and all rely on ants for protection. "If they don't associate with ants," he said, "they don't produce a call."

But there is an unsolved mystery: Some caterpillars associate with ants
but lack the sound-producing organs. Yet they also produce a call, and Dr.
DeVries has recorded it. "I don't know how they do it," he said.

All in all, he said, his findings provide "documentation of levels of complexity that I certainly never thought existed." And the ecological and evolutionary complexities of the ant-wasp-caterpillar-plant relationship, he
said, are a mere hint of broader-scale intricacies.

From Mexico to Venezuela, there may be 200 species of Riodinids—
no one knows exactly how many—that are symbiotic with ants. Habitats
change from north to south, and plant, ant and wasp species change with
them. In a given place, species vary with the seasons as well. This means
that there are innumerable possible combinations of symbiotic species of
plants and insects, each with a different set of evolutionary adaptations.

"It could get very complex quite quickly and involve an enormous
number of species," Dr. DeVries said. For a scientist trying to work through
this ecological and evolutionary thicket, he said, "it's pretty easy to get lost
real quick, trying to account for all those interactions.

"If we knew even a fraction of the life histories of all the Riodinidae
species," he said, "there could be many, many more fascinating stories than
singing caterpillars."

Given the rate at which tropical forests are being destroyed, he said,
the search is "a matter of some urgency."

—WILLIAM K. STEVENS, August 1991

Serenade of Color Woos
Pollinators to Flowers

MANY FLOWERING PLANTS woo their insect pollinators and gently direct them to their most fertile blossoms by changing the colors of individual flowers from one day to the next, a researcher has found. Through color cues, the plant seems to signal to the insect that it would be better off visiting one flower on its bush rather than another.

The particular hue tells the pollinator that the flower is likely to be far more engorged with nectar than are neighboring blooms; that nectar-rich flower also happens to be sexually ripe and ready to either dispense its pollen or to receive pollen the insect has picked up from another flower.

Thus, the color-coded communication system benefits both plant and insect and means that plants do not have to spend precious resources maintaining reservoirs of nectar in all their flowers; they simply advertise exactly where the bounty is.

"What's surprising about this is that plants really are playing a more active role than we had anticipated in getting pollinators to do just what they want them to be doing," said Martha R. Weiss of the University of California at Berkeley, who wrote the report. "The plant is saying to the insect, 'Go to this flower, not to that one.'" Her paper appears in the journal *Nature*.

For example, she said, in the lantana plant, an ornamental bush with tiny but abundant clusters of brilliant blossoms, a flower starts out on the first day of blossoming as yellow, when it is rich with nectar and pollen.

Influenced by some signal in the environment that Ms. Weiss has yet to identify, the flower begins changing from its innate yellow tone toward red by activating the production of a deep crimson pigment called anthrocyanin. It turns orange on the second day and red on the third, by which point it has no treats to offer insects and is no longer fertile.

But an abundance of brilliant blooms, whatever their color, is necessary to attract a pollinator from far away. Reaching the bush, the insect can use the variations in color to home in on fertile yellow specimens.

On any given lantana bush, only 10 to 15 percent of the blossoms are likely to be yellow and fecund. But in tests measuring the responsiveness of butterflies, Ms. Weiss discovered that the insects visited the rare yellow flowers at least 100 times more often than would be expected from random visitations.

Through experiments with paper flowers and painted flowers, she demonstrated that it was the color cue the butterflies were responding to, rather than, say, the scent of the nectar. Indeed, when she walked into a test area of her lab wearing a gray sweater with yellow specks on it, the butterflies eagerly fluttered to the specks.

Ms. Weiss said that while other researchers had observed color changes in a handful of flowering species before, few had realized how widespread such switching is. She demonstrated color shifts in 74 families of plants, with some changing from white to red, others from yellow to purple, etc.

She also showed that dozens of pollinating species of butterflies, bees and flies responded to those changes, always choosing colors and patterns indicating the likelihood of pollen—and nectar.

"The most spectacular aspect of Martha's work is the breadth of her survey," said Dr. Maureen Stanton of the University of California at Davis. "This shows that from an evolutionary standpoint, there must be very strong selection acting on plants to evolve floral aid signals to their pollinators."

—NATALIE ANGIER, November 1991

An Elusive Moth with a 15-Inch Tongue Should Be Out There

IT MAY NEVER HAVE BEEN SEEN by human eyes, and it certainly has never been described in the scientific literature, but somewhere in the dense tropical highlands of Madagascar there must flutter a moth with a tongue that measures 15 inches in length, a scientist says.

Speaking recently at an entomology meeting in Reno, Nevada, Dr. Gene Kritsky of the College of Mount St. Joseph in Ohio said that there was no alternative but to posit the existence of a giant moth with a six-inch wingspan and a longer proboscis than any yet seen on an insect. He made his prediction based on the existence of a rare orchid in Madagascar, a flower that could only be pollinated by a moth able to stick its tongue way down a tube where a pool of tempting nectar is hidden.

The tube of the exotic bloom, called *Angraecum longicalcar,* is 16 inches deep, and it hangs down from the blossom's stem like a whip; the nectar fills up about an inch on the bottom. Thus, said Dr. Kritsky, a moth would need to uncoil its proboscis to 15 inches to reach into the tube, called a nectary, suck up the juice and incidentally detach the orchid's pollen packet at the same time. Dr. Kritsky believes that the pollinating moth is a member of the sphinx moth family, and that it would be found if a diligent scientist willing to brave Madagascar's political chaos were to search for it.

"The orchid and its pollinator must have coevolved," said Dr. Kritsky. "The orchid could only survive if it had a moth pollinator."

The entomologist said he was doing nothing more than taking his cue from Charles Darwin, the great naturalist and occasional soothsayer. In 1862, while studying a Madagascan orchid, *Angraecum sesquipedale,* Darwin proposed that there must be a moth with an 11-inch tongue, able to dip into the flower's foot-long nectary and pollinate it. In 1903, two entomologists discovered a sphinx moth in Madagascar with a proboscis that matched Darwin's prophecy.

While rummaging through a natural history library last spring, Dr. Kritsky came upon a description of an orchid with an even longer nectary, and he decided it must have a pollinator with an even lengthier tongue. Other entomologists who heard his presentation agreed.

"It seems highly likely that another species of moth with a slightly longer proboscis is out there, able to pollinate that orchid," said Dr. Nathan M. Schiff, a research entomologist with ARSB Research Laboratory in Beltsville, Maryland, a Government agency. "Darwin got it right, and Gene Kritsky is just following in his footsteps."

Dr. Kritsky said that the orchid and its fertilizing insect probably evolved together over time, with the nectary getting a bit longer, and the moth's tongue following suit, until the relationship between the two became exclusive, a type of interdependence that suited both: the moth, because it could always be assured of a nectar drink from the orchid, which no other insect could penetrate; and the flower, because it could guarantee that its pollinator would be focusing its efforts on longicalcar and therefore would be likely to deliver fertilizing pollen from one bloom to another.

Dr. Kritsky said that the orchid discouraged small, scavenging insects like ants from crawling down its tube to gorge on its precious nectar with finger-like projections along its nectary lining that trap the insects.

To retrieve nectar from the flower, he said, a sphinx moth would have to settle on a lip at the edge of the skinny tube, and carefully uncoil its proboscis. As the moth probed, a pollen packet would shake loose from the orchid and stick to the moth. When the insect flew to another orchid, the packet would become detached and fertilize the recipient flower.

Dr. Kritsky said he has never seen the longicalcar orchid himself, but that the Cincinnati Museum of Natural History, where he works as an adjunct curator, is trying to get a sample. As for its fertilizing moth, because the insect is likely to be nocturnal, and because it must live in a remote region of the Madagascan jungle, where the longicalcar orchid blooms, it may never have been seen by a human being.

"Nathan Schiff told me I had to go find this thing," said Dr. Kritsky. "He said if I didn't, he would do it himself."

—Natalie Angier, January 1992

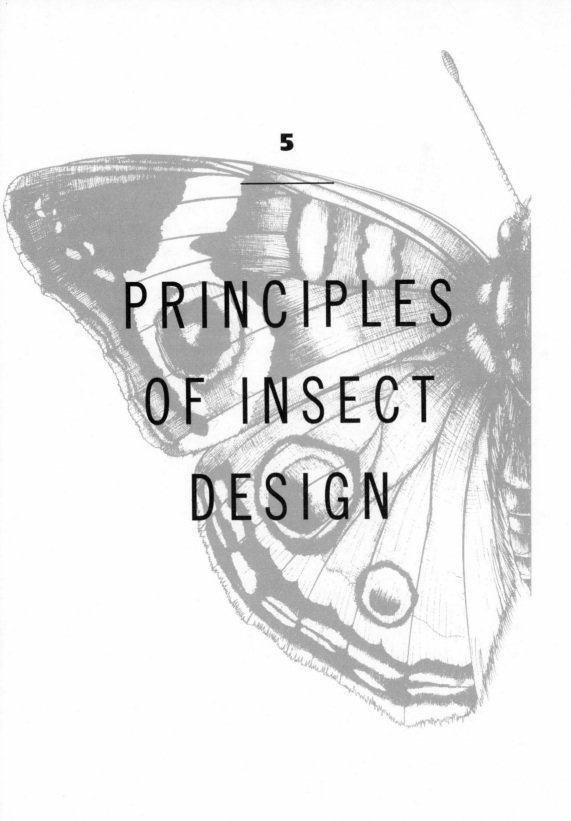

5

PRINCIPLES OF INSECT DESIGN

Ask an engineer to design a miniature robot that walks, flies, finds food, courts and reproduces itself, and he will explain that the first task may be possible but the rest are out of the question. The computing abilities of an insect's brain, small though it is, and its body's use of strong, lightweight materials, at present lie way beyond human imitation.

Insects possess many of the attributes of larger animals, such as complex social and courting behavior, but everything is compressed to a Lilliputian scale. Scientists are just beginning to understand the processes that go into designing an insect.

Engineers can describe the principles of insect flight, but the basic secrets of insect design lie in the genetic instructions imprinted in the DNA. Biologists have made a start in identifying some of the genes involved in creating one of nature's miracles, the painting of a butterfly's wing.

How Nature Makes a Butterfly's Wing

THE SPRAY OF COLORS on a butterfly's wing, a vivid pattern painted with the iridescent dust of thousands of tiny scales, ranks among nature's more entrancing mysteries. Though some like to keep wonders inviolate, others find a greater pleasure in understanding the elegant genetic machinery by which such beauty and diversity are generated. For the latter, an article in *Science* goes a long way toward explaining the miracle of how a butterfly's wing is made.

A team of biologists at the University of Wisconsin in Madison has traced out how a handful of genes becomes activated in the embryonic wing disks that are secretly budding while the butterfly is still a lumbering caterpillar. The genes produce substances that mark the position of each cell in the future wing with almost the precision of defining squares on a piece of graph paper. Once the coordinates are fixed, another crew of genes sketches out the dance of eyespots, chevrons, bands and dashes that make up butterfly wing patterns.

This soaring flight of understanding would have required decades of work, had not the main principles been brought to light over many years by those who study the development of the *Drosophila* fruit fly, a standard laboratory organism for biologists. In the work of just a few months, the Wisconsin biologists, led by Dr. Sean B. Carroll, used the powerful techniques of molecular biology to search for the butterfly's counterparts of the genes known to control the development of the fruit fly's wing. Although flies and butterflies diverged from a common ancestor some 200 million years ago, the DNA structure of their genes is still similar enough that fragments of fruit fly genes could be used to tag, through chemical affinity, their long lost cousins in butterfly cells.

The genes that control the development of the fruit fly's wing include those known as apterous, invected, scalloped, decapentaplegic, wingless

Drawing a Wing Pattern

A set of genes that control development of butterfly wings has been discovered in the buckeye. They lay down patterns on the embryonic wings while butterflies are caterpillars.

Butterfly

Caterpillar

Pupa

Wing Shape Defined by Gene

Gene called 'wingless' is switched on around edges of embryonic wing disk, making unneeded cells die away, leaving wing correctly shaped as if stamped out by a cookie cutter.

Color Development

Pattern laid down in caterpillar becomes colored as each wing cell produces scale of a single hue. Colors come from chemicals derived from the caterpillar's foodplant.

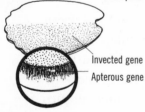

Invected gene

Apterous gene

Genes Define Coordinates

A gene called 'invected' is expressed in cells in the hindpart of each wing: the 'apterous' gene is switched on to mark cells as part of the wing's topside or underside.

From Simple Elements, a Profusion of Patterns

In the standard wing pattern shown here, each segment has the same basic elements. Butterflies vary the elements in each segment to make their overall wing patterns, creating a profusion of different patterns.

Michael Rothman

178

and distal-less. Apterous and invected, Dr. Carroll's team has found, serve as two of the positioning coordinates of the butterfly's wing. Apterous marks a cell as being part of the topside or underside of the wing; the signal is simply that apterous is switched on only in the topside cells, not in the underside. Similarly, the product of the invected gene is expressed only in the lower half of the insect's forewings and hindwings. There is probably a third gene, not yet identified, that tells cells how far out along the wing they lie from the insect's body.

In this way every cell's place is defined by how much of each positioning gene it expresses, so that an invisible genetic grid is laid across the wing. This grid is presumably the graph paper on which the pattern-drawing genes unveil the markings unique to each species.

The next act in the genetic drama is that the gene called wingless becomes active in cells around the edge of the embryonic wing disk. The gene is so named because in fruit flies it is essential for growing a wing (it is called wingless because when the gene is absent the fruit fly fails to form a wing). Butterflies have evolved a quite different use for it. The cells in which wingless is switched on receive a message to destroy themselves.

H. Frederik Nijhout, a Duke University biologist who has studied wing patterns for many years, recognized that the way to form the distinctive shape of a butterfly's wings would be to stamp them, as if with a cookie cutter, from the simple round disks in which they exist in the caterpillar, and that this could be arranged if the unneeded cells were somehow just to die away. Wingless is the magical cookie cutter. To sculpture the graceful streamers of the swallowtail butterfly or the scalloped edges of the tortoiseshell, nature trims away the surrounding cells by clicking on the wingless gene; its deadly molecular message, as soon as it is translated into its proteinaceous equivalent, bids the cell destroy itself.

After position and wing shape comes the imposition of pattern. The basic units of the wing are the wedge-shaped segments, bounded by veins, that run from the wing's root out to its side. Though each segment may bear a different pattern, all can be considered variants of the same basic theme, made up of four bands, an eyespot, a chevron shape and a final edge band.

Within each segment the eyespot is the major feature, and the Wisconsin biologists have found the gene that lays down the center of the eyespot. It is a gene called distal-less, which in fruit flies defines the wingtips and other extremities. In butterflies distal-less has been co-opted to create patterns. As the wing develops inside the maturing caterpillar, a streak of cells down the middle of each veined segment of the wing starts to make the product of the distal-less gene.

The streak then fades, leaving only a rosette of cells at the outer edge of the segment in which distal-less is active. These disks are positioned just where the eyespot forms on the adult's wing, and presumably lay down its foundation.

All these genetic programs unfold in the embryonic wing disks of the caterpillar. In the pupal stage, the patterned wing cells develop a rainbow of tones as each crafts a scale of a single hue. The rich palette of dyes in butterflies' wings are all derived from chemicals called flavonoids, which the insects cannot make themselves and must sequester from their food plants. The genes that direct the pupa's color development process have yet to be discovered.

Dr. Carroll and his colleagues worked with the buckeye butterfly, since it is easy to raise in the laboratory and since Dr. Nijhout had used it as the model for much of his analysis of butterfly wing patterns. They were pleased to find it used the same basic genes as the fruit fly, yet had developed some novel uses for them, Dr. Carroll said.

Though the action of these genes is far from the whole story, it goes a long way toward explaining the mechanics of how nature paints a butterfly's wing. As often happens in evolution's recipe book, a complex pattern is built up by reiterations of a simple theme. "Again and again what we are going to see is familiar genetic components being wired into similar programs," said Dr. Carroll. "There's a limited set of genes and proteins in the universe, and the way you generate diversity is by making small variations on these standard regulatory programs."

There are 12,000 living species of butterfly, each with its own distinctive decoration and most bearing quite different livery on their topside and underside wings. Yet this profusion of patterns can almost all be analyzed as variations on the same basic elements, Dr. Nijhout has shown in his book *The Development and Evolution of Butterfly Wing Patterns*. Presumably, the forces that produced this rich portfolio included the handiness of the eyespot in startling away predators and the need for camouflage.

Though the work of Dr. Carroll's team is only a beginning, it has excited biologists.

"It's very high quality and beautiful work," said Matthew P. Scott, a leading developmental biologist at Stanford University Medical School. "Its main importance from my perspective is that we're discovering the genes and molecules that control development are incredibly conserved across the animal kingdom, but since animals are very different, we have the problem of reconciling the conservation of the tool kit with the variation among species. Their work demonstrates both the conservation and that the tool kit can be deployed to create variations between species."

Dr. Nijhout said, "This is the first molecular and genetic handle we have on the process of pattern formation in butterflies." By building on the fruit fly work, the Wisconsin team has taken a leap of understanding, he said, since "it could have taken centuries before we would have gotten a gene like distal-less out of butterflies."

To those who would object that he has destroyed the mystery of the butterfly's wing by explaining it, Dr. Carroll replies that understanding how diversity is created from simple elements is something that "enriches our appreciation."

"Less than one tenth of 1 percent of species that have ever lived are still alive today," he added, "and wouldn't it be nice to understand creatures we can never see, like dinosaurs and the dragonflies with one-meter wingspans?"

—Nicholas Wade, July 1994

Now Playing at a Nearby Lab: "Revenge of the Fly People"

GODZILLA, *genital-less, gut feeling, gouty legs, goliath, gooseberry distal, ghost, glisten, gang-of-three.*

The quirky names, each carefully annotated, can be found on a Web site known as Flybase, the compendium of knowledge and home page of the world's 5,600 fly people. It is a point of honor among fly people, as biologists who study the laboratory fruit fly call themselves, to avoid giving unimaginative names to the new genes they identify in their favorite organism.

The fruit fly is a little speck of a thing, most noticeable in the form of the hovering clouds that emerge from that forgotten peach at the bottom of the fruit bowl. But this little speck has now become the center of a vast research enterprise, the focus of attention of many of the world's best biologists.

Their goal is to take the fly apart and put it back together by thorough analysis of its genes. When they can do that, they hope to understand several of life's deepest mysteries, like how a complex organism develops from a simple egg and how a nervous system is wired to produce a certain pattern of behavior.

Medical research agencies willingly support fruit fly research because its relevance to humans has turned out to be surprisingly direct. Many of the most important fruit fly genes, like those that tell the developing embryo to produce organs in certain places, have been found to have counterparts in humans. The fly and human versions of these genes are not identical, but have recognizably similar DNA sequences, reflecting their descent from a common ancestral gene some 600 million years ago.

Fly people are now positioned on the forefront of biological research. Since their organism is so easy to study, many significant genes are discovered first in flies, allowing other researchers to find counterparts in mice or men. The fly people's penchant for playful gene names has tended to cross over, too. A gene called fringe in the fruit fly has vertebrate counterparts named manic, lunatic and radical.

The fly's present preeminence in biology comes after a series of ups and downs. The *Drosophila* fruit fly was first chosen for scientific study in 1908 by Thomas Hunt Morgan of Columbia University. Morgan and his pupils worked out much of what is now standard genetics on the fruit fly. But *Drosophila* took a backseat after the discovery of DNA in 1953. The pioneers of modern molecular biology did their work in viruses and bacteria, the simplest forms of life. Fruit fly research stagnated.

The fly flew back into prominence when biologists skilled in the techniques developed for studying viruses and bacteria decided to tackle the far more complicated cells of higher organisms, and chose *Drosophila* as their vehicle. Several genetic techniques were developed in the fruit fly by Dr. David Hogness of Stanford University, and by students of his such as Dr. Gerald M. Rubin of the University of California at Berkeley.

By the 1980s, spectacular discoveries began to emerge about the genes that govern development of the early embryo. Some of these processes were special to insects, but the master genes that controlled them seemed to be part of a general plan for patterning all animal bodies.

Until the discovery of the patterning genes, said Dr. Ralph J. Greenspan of New York University, "it was very difficult to gain acceptance in other fields because they thought flies were irrelevant.

"Gene homology with humans really legitimized the fly," Dr. Greenspan said of the similarities in the genes' DNA sequence. *Drosophila* researchers who had sought federal grants on the grounds that their work would be relevant to humans found their arguments were even truer than they had thought, Dr. Rubin noted.

But at their moment of triumph, the fly people were threatened by rival tribes of biologist seeking knowledge from other totem animals. Some favored a tiny transparent worm, known as *C. elegans*. One of the fly people's stars, Dr. Christiane Nusslien-Volhard, now of the Max Planck Institute in Tübingen, Germany, began development of another model organism, the zebra fish of tropical aquariums. A technique for deleting genes put the mouse on the map as a workable model.

The National Institutes of Health supports research on all these animals as part of its human genome project, to which it deems each will contribute. For a time it seemed that creatures with backbones would provide a surer guide to human genetics. But the fly people have fended off their rivals, at least for the moment, saved by the surprising and extensive overlap of the genes among all the model organisms, and because the techniques for manipulating the fly's genetics are so much farther advanced.

Drosophila research is now flourishing as never before. In the late 1980s about 200 researchers attended the annual national fly meeting. At the recent meeting in Chicago, some 1,400 people showed up. Dr. Kathy A. Matthews of Indiana University keeps the Fort Knox of the fly world, a collection of some 4,500 different strains of *Drosophila* maintained at the stock center at Bloomington. She sends out some 600 shipments every week to fly laboratories, a quarter of them outside the United States.

"The field has been growing every year for the 10 years I have been here," Dr. Matthews said. "We have been growing by 20 to 30 percent a year."

The fly people compete with other biologists for money and graduate students but increasingly collaborate with them in finding counterpart genes. Still, *Drosophila* workers see themselves as a distinct community, glued together by shared institutions and values.

Their accumulating knowledge of fly genes, once broadcast in a mimeographed newsletter, is now stored in an electronic archive called Flybase (http://flybase.bio.indiana.edu, with mirror sites in Australia, Japan, the United Kingdom and Harvard). A separate data base, Flybrain, has been started for those who explore the genetics of the fly's nervous system (http://flybrain.ub.uni-freiburg.de). And the Berkeley *Drosophila* Genome Project, also supports a growing electronic archive (http://fruitfly.berkeley.edu).

The aim of the project is to work out the sequence of the 163 chemical letters of the fruit fly's DNA. Dr. Rubin, the project's director, said he hoped to complete the sequence within five years.

Fly people view themselves as being more open than other biologists and more generous with research materials. Their community has an ethos, which they attribute to Morgan, their founding father, of sharing fly strains and techniques even with close competitors. "It's that tradition—not a tradition you'd find in the tumor virus field, to name one—that has been tremendously powerful," said Michael Ashburner of Cambridge University, the author of a textbook known reverently as the Talmud of fly genetics.

In other fields, colleagues sharing a genetically altered mouse or a special reagent will often demand co-authorship on the resulting paper. The rule among the fly community is to share without obligation, Dr. Rubin said, a custom he views as conducive to rapid progress.

Yet giving out a strain or technique can quickly nullify a biologist's edge over competitors.

"You are virtually blacklisted if your don't share," said Sean B. Carroll of the University of Wisconsin at Madison. "I think everyone is more nervous, because there are so many extremely competent and hardworking people and your own advances are shared so quickly. You may make one long stride but it quickly becomes a level playing field until you may take your next step."

Besides competitive colleagues, another matter that constantly weighs on fly people's minds is virginity. Fly genetics depends on substituting the experimenter's strict designs for nature's unplanned shuffling of genes. Complicated, multigenerational breeding schemes can be ruined if the females enter into love matches instead of arranged marriages. So every morning and evening the fly breeder must collect newly hatched females before they are ready to mate.

"Arrange your crosses to compensate for nonvirginity wherever possible," Dr. Greenspan, of New York University, advises in *Fly Pushing*, a handbook of fly genetics (Cold Spring Harbor Laboratory). "This is especially important if your social life should suddenly become complicated and you miss the odd virgin collection."

Drosophila researchers sometimes call themselves fly pushers, because much of their time can be spent sorting anesthetized flies under a microscope. One morning in Dr. Greenspan's laboratory, two of his assistants, Nanci Kane and Susan Broughton, showed how the flies were stunned with a puff of carbon dioxide. After examination, appropriate partners were sucked gently into a glass tube and blown into a tiny nuptial chamber known as a mating wheel.

Fruit fly larvae are raised in small bottles and dine on a slush of baker's yeast, molasses and cornmeal. The passage from egghood to parenthood takes 10 to 14 days, depending on temperature. It is this rapid generation time along with numerous offspring and specially visible chromosomes, that makes the fly so convenient for geneticists.

Fly people, in turn, take good care of their minuscule charges. The flies often need cosseting, since strains may become enfeebled under a burden of weird mutations. Rather than grapple with agricultural customs regulations, some fearless fly aficionados are said to smuggle stocks across borders, strapping the bottles to their bodies to avoid the radiation of luggage-scanning machines. (The laboratory fruit fly is harmless, unlike its orchard-razing relative, the dreaded medfly). Two things keep fly people awake at night: mite infestations and incubator failure, either of which can wipe out precious stocks and months of crosses.

Flies have a delicate courtship in which the male sings by vibrating his wing; the female will accept him if she likes his song. Charming as the ritual may be, it is not completely spontaneous. The roles of both partners are governed by specific genes, a fact that has opened a new branch of inquiry. Most fly people still study the development of the embryo, but many are now using the fly to explore the genetics of behavior.

Drosophila geneticists have become so adept at designing genes to order that they talk of engineering the fly. But when asked whether he thinks of the fly as just a biological machine, Dr. Ashburner, of Cambridge, said: "No, I don't. We do engineer it like a machine, but I think most of us who work with flies realize just how bloody complex the thing is and how little we understand what is going on. The regulatory networks are very complex and we are only skimming the surface."

Dr. Greenspan, too, considers *Drosophila* more than a machine. One reason is that the fly's behavior is very varied, as is necessary to cope with the variability in its surroundings. He also admits to a certain feeling for the insect.

"There is a real sense of affection that people have for the fly," he said. "It is hard to account for. You always refer to it as 'the little fly.' In a sense it is a matter of respect for it, for what it has taught us and the idea that if you are appropriately humble in terms of what it can do for you, you can get an enormous amount out of it."

—Nicholas Wade, May 1997

A Gene by Any Other Name

A selection of *Drosophila* genes with colorful names and functions.

GENE	FULL NAME	DERIVATION AND EFFECT ON FLIES
SPDK	SPOTTED DICK	AFTER A BRITISH PUDDING OF THE SAME NAME; THE OVER-CONDENSED CHROMOSOMES IN BRAIN CELLS RESEMBLE THE FRUIT IN SPOTTED DICK.
STID	STALLED	FEMALES ARE LETHARGIC AND DO NOT GROOM THEM-SELVES PROPERLY; FLIES WALK SLOWLY AND TEND NOT TO GO ANYWHERE.
SHI	SHIBIRE	FROM A JAPANESE WORD THAT SUGGESTS WALKING AS IF IN A DRUNKEN STUPOR.
SINA	SEVEN IN ABSENTIE	FLIES ARE LACKING THE SEVENTH OF EIGHT PHOTORECEP-TOR CELLS THAT FORM EACH UNIT OF THE COMPOUND EYE.
SK	STUCK	MALES CAN GET THEIR SEX ORGANS IN BUT NOT OUT; THE PROBLEM AFFECTS SOME BUT NOT ALL MALES; THOSE THAT SUCCEED IN DISENGAGING HAVE APPENDAGES HELD IN ABERRANTLY PROTRUDING POSITIONS.

Aerodynamic Secrets of Insect Flight

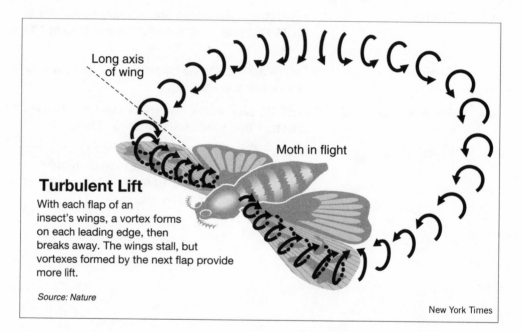

Long axis of wing

Moth in flight

Turbulent Lift

With each flap of an insect's wings, a vortex forms on each leading edge, then breaks away. The wings stall, but vortexes formed by the next flap provide more lift.

Source: Nature

New York Times

WITH THE HELP of a giant, smoke-spewing mechanical moth, researchers say they have finally figured out how insects fly.

That insects fly is obvious to everyone who has tried to enjoy a picnic in the midst of clouds of flies and gnats or anyone who has come too close to a beehive. But, aerodynamically speaking, these bugs should be on the ground competing with ants.

Now, researchers at Cambridge University in England say they have discovered how insects like bees pull their big bodies into the air with tiny wings in ways that defy conventional aerodynamic principles. When flying, they say, insects create whirling spirals of air above the front edges of their wings, providing extra lift.

Dr. Charles P. Ellington and his colleagues said they had found that as insects flapped their wings downward, some of the air that flowed over the wings rolled up along the entire front edge like a cylindrical window shade. This vortex is in the form of a conical spiral that grows as it sweeps along toward the wingtip, they said.

The spiral of air clings horizontally to the leading edge of the wing and slides down from the base to the tip before being replaced with another vortex generated by the next downward flap, they found. The vortexes extend back behind the insect from each wingtip and form a ring around the puff of air generated by the wing stroke, photographs show.

Dr. Ellington said in an interview that the vortex created a region of low pressure above the wing, which generated lift because of higher pressure under the wing pushing up. "Each vortex is a low-pressure area that kind of sucks the wing upward, creating an extra lifting force," he said.

Scientists have long noted that the wings of many insects produce more lift than can be explained by traditional aerodynamic laws, but the source of the extra lift had remained unknown.

"Most people suspected that a leading-edge vortex would account for the extra lift, but studies in the past either didn't see the vortex, or the forces that they detected were smaller than expected," Dr. Ellington said.

The Cambridge group, in a report published in the journal *Nature,* said it had refined detection techniques by flying large hawkmoths in a wind tunnel and taking high-speed, three-dimensional pictures of smoke trails moving over their wings. In addition, the researchers were able to visualize the leading-edge vortex using a mechanical moth with a three-foot wingspan that shot smoke from the edge of the wing.

"Ellington's group has found a big piece of the puzzle of how insects fly by generating lift greater than their body weight," said Dr. Steven Vogel, a professor of zoology at Duke University who is a leader in studying the motion of animals. "While this doesn't answer all of our questions about insect flight," he said in an interview, "Ellington has put his finger on perhaps the single most important piece remaining to be described."

Professor R. McNeill Alexander, an expert in biomechanics at the University of Leeds in England, praised the work in a commentary published in the same issue of *Nature.* Dr. Ellington and his colleagues, he said, "give us the first clear evidence that one particular mechanism is the most important one." In effect, he said, "they have discovered how a typical insect flies."

Dr. Alexander said in a telephone interview that the biggest surprise of the work was the discovery that the leading-edge vortexes spiraled out along the wings toward the tips in a corkscrew fashion. This motion, he said, may help stabilize the vortex, delaying its separation from the wing and enabling it to retain its lifting strength longer than otherwise possible.

"It remains to be discovered whether similar effects occur in the flapping flights of birds," Dr. Alexander observed in his commentary. "In some cases, such as in slow or hovering flight, birds also generate more lift than conventional aerodynamics can explain."

To study insect flight, the Cambridge group focuses on hawkmoths because they are big, having wingspans of about four inches and flap their wings relatively slowly, making them easier to study. The researchers said the moth's wing movement was also typical of many insects.

In a series of experiments, hawkmoths were held in a loop of thread attached to a stiff wire to keep them in the breeze from a small wind tunnel. The blowing air contained thin streams of smoke to show how the air flowed over the wings and the researchers used a special video camera to capture the motion at 1,000 pictures a second.

Evidence of the wing vortexes appeared above and just behind the front edge of the wing when the moth flapped down. Instead of just flowing over the top, the scientists reported, a layer of air curled up into a cylinder along the edge.

The researchers saw what they thought were vortexes, Dr. Ellington said, but the effect was so small along the moths' wings that the smoke just flowed around it without revealing any details. At this stage, the scientists decided to build their monster moth, which they called "the flapper."

They made a computer-controlled mechanical moth whose body houses four motors and a gearbox to drive the movements of the robot wings. The wings, spanning three feet, are made of an elastic cloth stretched over a framework of brass tubes and joints, and are designed to generate smoke from the leading edge.

Although the mechanical moth flaps more slowly than the real thing, it accurately models the motion of insect wings, Dr. Ellington said. The machine was able to generate smoke directly into the vortexes formed, allowing scientists to see the complex patterns of spirals and loops moving from the wings for the first time, he said.

"The mechanical moth was the key," he said. "It allowed us to see the unexpected movements of the vortexes."

The researchers found that as the wings moved forward, a leading-edge vortex formed with each flap, producing moments of maximum lift before breaking away. This sudden loss of lift stalls the insect wing, much as happens with an aircraft wing that loses lift if its angle into the wind gets too steep, and it could fall if the next vortex did not follow on the next flap to produce another lifting force.

"This delayed stall mechanism probably is the primary means that insects use to stay in the air," said Dr. Michael H. Dickinson of the University of California at Berkeley, an assistant professor of integrative biology who specializes in insect movement. He praised the Cambridge work as "a beautiful analysis of the forces at work in flight."

But Dr. Dickinson said there was still much mystery left in insect flight and researchers still do not understand how insects control themselves in the air.

"Insects don't just stay in the air," he said, "they perform aerial maneuvers; move up, down and sideways; respond to changes in wind speed and direction. We don't know what the animal is doing to control these forces and it's going to take a lot more work by scientists with different specialties to find out."

—WARREN E. LEARY, December 1996

What's New About a Spider's Web?
It Takes a Computer to Know

USING A COMPUTER MODEL to study the behavior of spiderwebs, researchers say they have discovered the solution to what one of them described as a crucial problem for the spider—how to stop a relatively massive insect moving at a fast speed.

If the web were too stiff, said Lorraine H. Lin, a structural engineer and an author of a new study on spiderweb dynamics, "the insect would bounce back out like hitting a trampoline," adding, "If there was too much give, the threads would break."

As a fast-moving insect hits the web, a fine net stretching over a relatively large area, the impact stretches and deforms its fibers, the researchers said. The drag produced by the movement of the web against the air helps to dissipate the energy of the blow.

This aerodynamic resistance improves the chance of the web's staying intact by quickly damping vibrations and increases the likelihood of the insect's sticking to the web long enough for the spider to reach it for the kill.

The scientists, at Oxford University in England, said this property of the circular, or orb, webs of a common garden spider, *Araneus diadematus,* also caused the web to billow like a sail in a breeze, a factor that increases its chance of being destroyed in a high wind. Taking this danger into account, they said, some spiders apparently try to control the benefits and risks of the web's air resistance by building it at an angle to the primary direction of the prevailing wind.

In a study published in the journal *Nature,* Dr. Fritz Vollrath and Dr. Donald T. Edmonds of Oxford joined with Ms. Lin, a structural engineer with Ove Arup & Partners, an international engineering construction concern in London, to examine the orb spider's web. The researchers built a computer model of a web using a customized version of a structural analysis program that is mainly used for automobile crash simulations.

The computer simulations indicated that while the stiff supporting threads of a web dissipated energy from the impact of an insect, this reaction was not enough to account for all the energy absorbed by the web. Only by including the air resistance, or drag, caused by the rest of the net could the behavior of real webs be reproduced, the report said.

The researchers said they confirmed their computer findings by firing Styrofoam bullets at real webs, including some with sections of the web net cut away to reduce aerodynamic damping that disperses energy.

Dr. George W. Uetz of the University of Cincinnati, a biologist who specializes in spider behavior, said the new research was significant because "they are applying new technology to an ancient, biological structure and coming up with interesting, unexpected results." Since the web's support threads also transfer vibrations to the spider so it can detect prey, he said, a damping effect that localizes vibrations might help the spider locate its potential meal more quickly.

Experts believe that spider silk, thin water-soluble threads with a breaking strength greater than steel, first evolved 400 million years ago and orb webs emerged 180 million years ago as the creatures left the ground to hunt in bushes and trees. Research is increasingly showing that these webs, far from being simple nets that snare passing insects, are dynamic structures that respond to their immediate environment and even embody characteristics to attract prey.

Ms. Lin said in a telephone interview that spiders created orb webs by making two types of thread. They make a framework of dry, radial threads that radiate from the center like spokes on a wheel, and overlay these with a spiral of flexible, wet, glue-covered strands. While the radial threads are taut and under a slight tension when the web is in its resting state, the spiral strands are elastic and expand or contract their length as conditions change.

The glue on the spirals not only helps entangle insects with its stickiness, but also absorbs ambient moisture, she said. This water forms droplets along the thread and the surface tension of the water in the droplets pulls the elastic strand into little bundles to keep it from sagging when not under pressure. This droplet mechanism rapidly reels the thread in and out to keep it straight, depending on the load pulling on it.

The researchers found that the presence of these glue droplets increased the wind resistance of the web and enhanced aerodynamic damping by about 35 percent, she said.

"The web is a very dynamic structure made up of two types of thread, one that is super elastic and one that isn't," Ms. Lin said, and together they solve the spider's crucial problem.

—WARREN E. LEARY, January 1995

6

THE
UNWELCOME
SIDE OF
INSECTS

Of the millions of existing species, only a handful ever become pests, often when humans have upset the balance of nature, or transplanted an insect to a strange home where the natural enemies that usually keep its population in check are lacking.

Most insects have a reproductive strategy the opposite of humans; instead of investing a lot in a few offspring, they invest a little in many. Thus when conditions are right, insect populations explode. These infestations range from the merely annoying, like the congregation of Halloween bugs, to threatening plagues like locust swarms.

DDT and the chemicals used to wage warfare on insect pests were at first spectacularly successful and saved many lives. But the chemicals destroy beneficial insects too and leave harmful residues; and insects in any case develop resistance to them. These problems have prompted development of biological methods of control, such as the insect-destroying bacterium *Bacillus thuringiensis*. But insects have started to develop resistance to the *Bacillus* toxins as well. The war against insect pests still has far to go.

Mystery Bug Infestations?
Call in the Insect Sleuth

THE LARGE UPSTATE New York hospital was desperate for help. Flies appearing from seemingly nowhere had invaded the surgical suites, dropping onto patients and operating-room personnel in the midst of surgery. With the hospital unable to stem the invasion, the whole surgical staff walked out in disgust and patients in need of operations had to be transferred elsewhere.

The administrator put in an emergency call to Dr. Edgar Raffensperger, a professor emeritus at Cornell University with an unusual specialty. Dr. Raffensperger is an insect sleuth, a cultural entomologist who is an expert on the role that insects, spiders and other arthropods play in people's lives, like it or not. The hospital, needless to say, did not like the flies and asked Dr. Raffensperger to figure out where they were coming from and how to get rid of them—fast.

"It was elementary, my dear," the professor said in an interview. Over the years the renowned bug detective has solved any number of arthropod mysteries, like the case of the impeccably clean factory that kept finding flour beetles in its jars of baby food, or the problem of the beer-bottle warehouse invaded by hordes of black widow spiders, or the strange affair of the flies that closed down an airport by disrupting the electronic gear in its control tower.

The first trick of his trade, Dr. Raffensperger explained, is a precise taxonomic identification. In the case of the upstate New York hospital, he recognized the invaders of the surgical suite as cluster flies, which are often seen on cool summer and early autumn days lazing on sunlit walls.

The next principle, he noted, requires intimate knowledge of the suspect's biology. Like other cold-blooded insects, cluster flies rely on the environment to warm them. When the temperature exceeds 53 degrees

Fahrenheit, the flies seek out the sun, but when it gets cooler, they prepare for hibernation. They will crawl into buildings through any hole they can find—a weep hole in the masonry, a crack in the mortar or a tiny hole in the putty of a sash-cord window.

Applying this knowledge to the New York hospital, "Detective" Raffensperger quickly understood what was prompting the invasion of the surgical suites. The flies, he determined, were using the hospital for a winter den, entering the building through tiny holes in the brick walls. When the bright lights were turned on in the operating rooms, warming the air, the flies emerged from hibernation and crawled into the sterile rooms through tiny openings, like the light sockets. The light-seeking flies would bang into the hot fixtures and drop onto the patient, doctors and nurses below.

"Since the flies hibernate in secluded places, there's no way to get to them, short of tearing down the whole building," the entomologist noted. Nor could the hospital find and seal every hole a fly could conceivably get through.

"A more practical alternative," Dr. Raffensperger said, "is to spray the entire outside of the building with an insecticide that would keep the flies from getting inside in the first place." The entomologist recommends a long-acting pesticide that is minimally toxic to mammals: synthetic forms of pyrethrin called cypermethrin or cyfluthrin, which will keep the flies away for months.

Cluster flies also proved to be the culprits in the case of the Rochester International Airport in upstate New York, where the flies got into the air controllers' equipment and forced a temporary shutdown.

Since the incident at the upstate New York surgical suite, Dr. Raffensperger has received similar distress calls from many other hospitals, from Nova Scotia to Washington State. Although Dr. Raffensperger, like a well-seasoned physician, can often make a telephone diagnosis and prescribe an effective remedy over the phone, the "patient" usually wants him to come to the house.

A house call was unavoidable in another operating-room invasion by tiny fungus gnats, insects smaller than one sixteenth of an inch that live on rotting vegetable matter. The gnats got into the operating room through the filter system meant to transport especially clean air, he found. Since the filter was on the roof of the building, the insect sleuth headed straight for the crawl space under the hospital's roof. It took him half a day of searching on his hands and knees with a flashlight to find the rotting wood that supported the duct work of the filtration system and the population of gnats.

"There they were, breeding like crazy," he said. The gnats were being sucked out of the crawl space by the filtration system and fan and deposited in the operating room, which, paradoxically, had been designed to maintain positive air pressure to keep contaminated air from outside rooms and halls from flowing in.

The hospital had no choice but to replace its entire roof, removing all the wood and rebuilding it with materials that would not attract fungi and, in turn, fungus-loving gnats.

On another occasion Dr. Raffensperger was called to a West Coast beer-bottle warehouse that had been taken over by black widow spiders. The terrified workers refused to enter the warehouse or handle the cartons of bottles because they were covered with the fearsome spiders. The brewery had been forced to shut down for lack of bottles.

"I went into the warehouse, saw a carton with four or five spiders on it, scooped them up and put them in my hand as the workers shrank back in terror," he recalled, his eyes dancing gleefully at the memory. "Despite their deadly reputation, I knew that black widows are really very docile. They will not attack unless they are attacked. There are only six recorded deaths from black widows, and they all occurred in children who were bitten in privies after they inadvertently sat down on the spiders," he said.

Why had the spiders infested the warehouse? "The area surrounding the building was very hot and dry, and the spiders sought refuge through the open doors of the warehouse, which was cooler and damper," he explained, his tone implying that even Dr. Watson could have figured that one out.

The solution was also simple: kill all the spiders in the warehouse with a short-acting pyrethrin and spray a longer-lasting barrier of pesticide around the building to prevent new infestations.

The case of the beetle-infested food jars presented a different problem. A spotless processing plant received newly made, and presumably clean, glass jars from a supplier, blew the jars out to get rid of any paper dust, filled them with food and sealed them. So how come tiny beetles were showing up in jars of food sold to consumers?

Dr. Raffensperger traced the problem to the glue used to seal the cartons of jars, which were stored in an unheated warehouse before they were filled. Water condensed on the cold concrete floor of the warehouse and moistened the glue. Fungi grew in the damp glue, which attracted fungi-eating beetles; some of the beetles dropped into the clean jars and resisted leaving.

To prevent recidivism of the culprit, Dr. Raffensperger recommended that heaters and fans be installed to control moisture in the warehouse.

Another case of beetle infestation nearly stumped him. How on Earth, he wondered, were flour beetles getting into jars of baby food produced and packaged in an immaculate factory? Flour beetles are drawn to grains and flour, but they were not being bottled in the processing plant.

Then he discovered that the beetles were dropping from the ceiling into the food as it was being bottled. But why were they on the ceiling in the first place?

Looking outside, Dr. Raffensperger noticed a defunct flour mill next door. Flour dust from the mill had settled on the roof, and the beetles were thriving on the fallout. Since the roofing was not airtight, some of the beetles managed to crawl through the ceiling and fall into the food.

Dr. Raffensperger has also served as a forensic entomologist, establishing the time of death "within two hours" after checking the status of carrion-loving insects. "The police will bring me maggots from a body they found and ask how long ago the person died. All I need to know is the location and the temperature and weather records for that location, and I can tell even more accurately than a medical examiner."

Flies with flesh-eating larvae deposit their eggs within two hours after an animal dies, as long as part of the day is above 60 degrees Fahrenheit, the entomologist explained. These larvae, or maggots, go through three stages, which vary in length from species to species, before becoming pupae 48 hours later.

"If you know the species of fly, you can tell from the stage of the maggot's development how long ago it was laid on the corpse," he explained, his voice and eyes trailing off as if he had been suddenly struck by an inspiration, perhaps to write *The Maggot Murder Mystery*.

—JANE E. BRODY, August 1992

Tiny Predator Imperils the Eastern Hemlock

MORE AND MORE, it looks as if the Eastern hemlock, with its delicate evergreen foliage, its graceful lines, pyramidal crown and branches that droop nearly to the ground, may go the way of the chestnut and the elm.

Disease imported from abroad all but wiped out the latter, two of North America's most impressive and cherished trees, earlier in this century. The hemlock has long been under attack by an imported insect that sucks out its juices, and the insect's steady northward advance has lately been provoking concern in Connecticut, New Jersey and southern New York.

Now, for the first time, the insect is approaching the hemlock's heartland in northern New England and eastern Canada, where it will be less impeded by gaps in the forest caused by urban development and where a vast store of food stretches out before it.

Although scientists are about to embark on a new effort to identify a natural enemy that might counteract the pest, they have so far found nothing to stop it from eventually destroying the Eastern hemlock throughout its range in the wild.

The killer is the woolly adelgid (pronounced uh-DEL-jid), a tiny piercing-sucking insect whose presence on hemlock branches is vividly announced by strings of small, fuzzy white balls—its egg sacs—that look much like artificial snow on a Christmas tree. Once feeding adelgids attack a tree, it can die within a year and almost invariably succumbs within four.

If the adelgid wipes out the hemlock in the wild, many Northeastern watersheds could be exposed to erosion and some river ecosystems could be vulnerable to disruptive and possibly destructive changes.

"The outlook is grim," said Dr. Mark S. McClure of the Connecticut Agricultural Experiment Station in Windsor, a leading researcher on the subject. He described the findings at a symposium on Northeastern forests at the New York Botanical Garden.

Potentially, he said in a subsequent interview, the adelgid can do to the hemlocks what chestnut blight and Dutch elm disease did to those two trees. "I see no resistance" to the woolly adelgid, he said. "The insect has the capability of destroying all sizes and ages of trees, and the tree just doesn't have time to respond genetically." Furthermore, he said, the adelgid's 30-mile-a-year rate of advance is "very fast" for an insect.

As the size of the infested area increases, the adelgid population expands and invades more and more areas, and now that it is approaching the hemlock heartland, the onslaught is expected to accelerate.

So far, since its discovery in eastern Virginia 30 years ago, the adelgid has spread throughout most of the Washington-Boston corridor, westward into Pennsylvania, Maryland and interior Virginia, northwestward into lower New York State and north into western and central Massachusetts, where it is approaching New Hampshire and Vermont.

The Eastern hemlock, which reaches a height of 60 to 70 feet and can live 600 years, grows in small, pure groves or in mixed stands with hardwoods. It lives under very precise conditions: moist, shallow soil, on steep embankments, in areas where shade tolerance is a virtue. It tends to dominate steep, northward-facing slopes, where sunlight is less strong.

"No other evergreens fill that bill," said Dr. McClure. The adelgid assault, he said, "opens up the potential for quite a bit of environmental change." Once the hemlocks were gone, there would be a lot of erosion and the temperature of rivers would rise because of a loss of shade. "It's hard for other trees to fill that gap," he said.

The adelgid is believed to have come to the United States from Japan by an unknown mechanism and route. It appeared first in the Pacific Northwest, where it adopted the Western hemlock as a host. As was the case in Japan, where the adelgid's host is the spruce tree, the Western hemlock proved resistant to the insects, and they did not proliferate.

Such is not the case with the Eastern hemlock. Nor is there any known natural enemy that might control the adelgid.

"We will be looking for a natural enemy in Japan," said Dr. McClure. "But none are known there either, though that just may mean that nobody's gone out and looked for it." If none is found, he said, "then the outlook isn't very good in terms of the forest."

Spraying the forest, as is done, for instance, to control gypsy moths, is impractical for the adelgids because the necessary saturation is not attainable with aerial spraying. With the gypsy moth, saturation is not necessary because the gypsy caterpillars move around so much that they are likely to contact the insecticide at some point. But the adelgids pretty much stay in one spot; those that are not hit by the spray are not harmed.

Dr. McClure held out much more hope for ornamental hemlocks, which are prized not least because they can be sheared, trimmed and shaped or left in their natural graceful state. Saturation spraying or injecting pesticides are practical alternatives for these trees because of their limited numbers. There are also some horticultural oils and soaps that kill the adelgids but are relatively harmless to other organisms.

Woolly adelgids are dormant in warm months and do most of their feeding from November through March. But Dr. McClure said that they can be successfully attacked with insecticide at any time of the year, at any stage of life.

In fighting other pests, it is common practice to fertilize the tree liberally so it can build up strength and mass to better withstand the damage inflicted by chewing insects. But with the woolly adelgid, said Dr. McClure, this only makes the problem worse. In experiments conducted by Dr. McClure, adelgids were three times thicker on fertilized trees than on unfertilized ones.

Piercing-sucking insects directly use nitrogen, the main component of fertilizer, for nourishment. "Presumably," Dr. McClure said, "they intercept the nitrogen taken up by the tree before the tree can make use of it."

—WILLIAM K. STEVENS, November 1991

From Boon to Bane
The Halloween ladybug, *Harmonia axyridis,* orange and black rather than red and black, is a thriving immigrant species. Biteless and disease-free, it eats aphids that attack corps, but it swarms in sheltered places, alarming homeowners.

Michael Rothman

Swarming Halloween Bugs
Haunt Humans

THE HALLOWEEN LADYBUGS are swarming—in numbers suitable for a Hitchcock film. They came from Asia, to Louisiana, to Georgia, and now they are in New York. In fact, throughout the Northeast, houses have been invaded by hundreds or thousands of the orange and black pill-size beetles, causing the human inhabitants to rush to their telephones in alarm.

Since mid-October, the Cornell Cooperative Extension office in Millbrook, New York (a sort of 911 for ecological emergencies), has fielded several hundred calls about ladybug swarms outside and inside Dutchess County homes. The Massachusetts Audubon Society's wildlife sanctuary in Lenox has also been deluged with calls. The bugs have been sighted as far north as Maine. There are even reports of a true catastrophe for a New England autumn—guests have threatened to leave bed-and-breakfast inns because of the beetles.

The beetles have triggered smoke alarms, crawled into bedding and drawers of clothing, and squeezed beneath the glare screen of computer monitors. They have invaded high-rise office buildings, restaurants and food-processing plants.

The ladybugs have literally covered the sunny side of white or light-colored houses, pelting windows like a sleet storm. Some homeowners say they have been traumatized by the experience. Others have found the situation simply frustrating. "One woman told me her new white curtains were ruined," said Stephanie Mallozzi, a Cooperative Extension horticulturist. "She lives in a log home and I guess the ladybugs thought it was a huge hollow tree where they could spend the winter."

The bugs are benevolent, not harmful. Scientists are quick to point out that Halloween ladybugs, so called because of their color and time of appearance, neither bite nor sting. They are not poisonous and carry no diseases. They cannot breed indoors like fleas or cockroaches, and they do not eat wood like termites. The ladybug explosion, entomologists emphasize, is a bonanza for fruit growers and foresters because these arboreal predators have a huge appetite and feed on aphids, which attack a wide variety of trees, including apples, peaches, plums, pines, oaks, maples, magnolias and tulip poplars. A single larva, which resembles a tiny orange and black alligator, can eat 300 aphids in two weeks before it enters the pupal stage and meta-morphoses into an adult.

But entomologists say that most Americans have an aversion to insects. In a survey in Kentucky, where the beetle made its debut in 1992, 4 of 10 homeowners said they would take action if they found a single ladybug in their house. And sometimes the insect's numbers are uncountable. "If woman wakes up in the morning and finds 20,000 ladybugs clustered in a corner of her kitchen ceiling, she doesn't care that they eat aphids," said Robert Norton, spokesman for the Agricultural Research Service of the Federal Department of Agriculture in Greenbelt, Maryland.

Also, the bugs do have an unpleasant side. When stressed, they exude orange blood that is both sticky and pungent. "If you go after them with a broom, you'll have orange spots all over your walls and the stuff is pretty evil-smelling," warned Dr. William H. Day, an entomologist with the Agricultural Research Service. The secretion, which oozes from joints in the legs, is the beetle's defense against such predators as birds. "They won't try to eat a second ladybug," Dr. Day said.

The name "Ladybug" dates to the Middle Ages when the beetles rid grapevines of insect pests and were dedicated by grateful friars to "Our Lady, the Virgin Mary." The Halloween ladybug, an Asian species known to scientists as *Harmonia axyridis,* is one of more than 4,000 described species of ladybugs in the world, including some 475 species in North America. Countless more probably await discovery in the tropics.

Harmonia axyridis has a checkered history in this country, and how it landed in the Northeast is far from clear. Nonnative ladybugs have been released in the United States since the late 1880s to prey on agricultural pests, accidentally imported insects with no natural enemies in the new world. The Halloween ladybug has been used in various biological control experiments since 1916. In more recent times it has been introduced in several states, including Connecticut, Louisiana, Maryland, Mississippi, Ohio, Pennsylvania, Texas and Washington. In Georgia, for example, 88,000 of the ladybugs were raised and freed between 1978 and 1981 in an effort to combat aphids found on pecan trees.

The beetle apparently had problems adapting to a new home, however, and Agriculture Department experts insist that none of those introductions resulted in populations of this species being established in the United States.

"The ancestors of the ladybugs we're seeing in such huge numbers probably arrived by boat in time-honored immigrant fashion," said Dr. Day, who works at the Beneficial Insects Introduction Research Laboratory in Newark, Delaware. He said that scientists from Louisiana State University found the first breeding population of immigrant *Harmonia axyridis* in 1988 in St. Tammany Parish, near the Port of New Orleans.

It would be easy for ladybugs clustered in a protected nook on a cargo ship to escape detection by port inspectors, he said. Four other Old World ladybugs have made the Eastern United States their adopted home since the 1960s, he said, and all were first found near seaports. None of the four, however, has become a public nuisance.

By 1990, the Louisiana immigrants seemed to have spread, and *Harmonia axyridis* became established at last in Georgia. Presumably the ladybug that had failed when it was introduced now thrived when it moved on its own. In 1992, one way or another, the beetle was finally found again on the Agricultural Research Service's Southeastern Fruit and Nut Tree Research Laboratory, near Macon, where it was first identified by Louis Tedders, an entomologist.

"Since then they have whipped up the East Coast on wind currents all the way to Maine and Quebec," Dr. Day said. The Halloween ladybug was first seen in New York near Elmira in February 1994, said Richard Hoebeke, an entomologist at Cornell University, and it conquered the entire state in less than two years. Mr. Hoebeke described the ladybug as oval-shaped, measuring one quarter of an inch long and three sixteenths of an inch wide. The beetle's hard wing covers (or elytra), which lift forward like hoods on some luxury automobiles so the insect's long flight wings can unfold, typically are yellow-orange with as many as 20 black spots. The pronotum, a plate between the ladybug's head and the elytra, is dirty-white with a distinctive black mark in the shape of the letter "M."

The insects' color as well as the number and size of their spots is highly variable, however. Some individuals have only a few pinprick spots, or none. Some specimens are brick red and others are black with orange blotches. As to the name of the newcomer, the Entomological Society of America has proposed "multicolored Asian lady beetle" as the official common name. "That's ridiculous," said Dr. Day. "They should choose something simple. Of course," he added, "a lot of people are calling it unprintable names." The reason people notice *Harmonia axyridis* is the species' unusual habit of aggregating in sheltered places, which often means occupied buildings, in October and November. The vast majority of ladybug species hibernate outdoors beneath leaf litter, under the loose bark of trees or in clumps of grass. But insulated attics, root cellars and cathedral ceilings are favorite overwintering places, and because these beetles are forest inhabitants, homes in wooded areas are particularly susceptible to infestation.

The tendency of the Halloween ladybug to congregate on white houses, Dr. Day explained, may be due to their ancient habit of wintering in crevices of limestone cliffs in Japan and elsewhere in eastern Asia. "People could paint their houses a darker color, but that's an expensive and not very desirable solution," he said. Caulking or weather-stripping points of entry such as loose-fitting window casings will keep the beetles out of modern homes, the scientist said, but owners of older houses could face more serious problems.

"If you've got hundreds of ladybugs in your home, imagine where the heat is going in winter," said Cornell's Mr. Hoebeke. Beetles trapped in warm living quarters will not survive, he added, but ladybugs that find cool hiding places will emerge on bright days in February and March. This time they will be looking for a way out of the house. Experts advise removing indoor aggregations with a vacuum cleaner, using a fresh bag and storing the beetles in a cool place for release in the spring.

The Halloween ladybug is not likely to go away. Females produce several generations a year, laying 20 eggs in a day, and the adult beetles can live two or three years. Mr. Hoebeke is concerned about the possible impact of this prolific alien on its native relatives. He believes that the nine-spotted lady beetle, once the best-known species of ladybug in the Northeast and New York's official state insect, has been wiped out in many areas by the large and aggressive seven-spotted lady beetle from Eurasia. "I won't say it's extinct," he said, "but intensive surveys have been done and the beetle hasn't been seen in the region since 1989."

Dr. Natalia Vandenberg, a Smithsonian Institution beetle specialist, agreed: "We're likely to see some displacement of native species with each exotic insect that's introduced, but it's unlikely that large ladybugs with wide distribution will become extinct.

"These days," she stressed, "scientists introduce insect predators that are very host-specific and have restricted host ranges" rather than generalists like this Asian species.

And nature may lend a helping hand in the form of parasitic flies and wasps that will check the population explosion of *Harmonia axyridis*. In Georgia, Mr. Tedders reports that "natural controls are happening right now and beetle numbers are way down from last year."

Entomologists continue to experiment, meanwhile, with other foreign ladybugs as biological control agents. Earlier this year, at the Connecticut Agriculture Research Station in Windsor, Dr. Mark S. McClure introduced a spotless black species "the size of a poppy seed" from Japan as a potential enemy of the woolly adelgid that is devastating stands of hemlocks in the East. "It hasn't been through a winter, so we don't know if it will become established," Dr. McClure said.

Ten years ago, Dr. McClure released "a thousand or so" *Harmonia axyridis* beetles in Connecticut as a control for red pine scale, which reached the United States on infected nursery stock shipped from Japan for the 1939 World's Fair in New York City. "The ladybug was a very effective predator on that scale in Japan, but our native red pine has heavily textured bark and the insect was able to hide throughout much of its life cycle. Without any food, the beetles left the stand and were never seen again."

Halloween ladybugs, of course, are now spotted throughout Connecticut. Are they descendants of those lost beetles?

"I hope not," Dr. McClure said.

—LES LINE, October 1995

Mite that Causes an Overwhelming Itch

FOR A 90-YEAR-OLD WOMAN, the itching that began a week after she underwent major surgery in a hospital in New York City was maddening. The itching was most severe at night, and she could not sleep. She became depressed, lost her appetite, and her skin reddened and oozed from constant scratching of a rash.

The doctors were baffled. Initially a dermatologist on the medical team suspected the itching was an unwanted reaction to a drug, and the doctor stopped all the ones she was receiving. Still the itching persisted.

Someone suspected an allergic reaction to the detergent used in washing the sheets. But common household lotions afforded no relief.

Six weeks after surgery, 30 pounds lighter and bone thin, the woman went home, still scratching. She went to another dermatologist who snipped a small piece of skin. When he examined the biopsy under a microscope, the diagnosis was a surprise: scabies.

The scabies cleared with application of a standard treatment for the condition.

Scabies is caused by a mite, *Sarcoptes scabiei*, that is one sixtieth of an inch long and is responsible for an epidemic in many countries. The mite is usually transmitted directly from one person to another from contact as intimate as sexual intercourse. The mite can also spread through contaminated clothing and bedding, which is possibly how the woman got it in the hospital in New York City.

Scabies produces a rash that can be easily confused with many other common conditions, like eczema, drug reactions, insect bites and infections.

Many dermatologists have made their reputations by diagnosing cases of scabies their colleagues had missed.

In seeking the cause of a rash, a doctor asks about other cases, whether it itches, what factors aggravate and relieve it and whether the severity varies with the time of day. For instance, one clue to scabies is that the itching tends to be worse at night.

A doctor also pays close attention to the sites where the rash appears on the body.

In adults, for example, scabies tends to occur most often in skin between the fingers, on the wrists, in the armpits, the elbows, breasts, genital areas, navel, buttocks and toes. Scabies generally spares the upper back, neck, face, scalp, palms and soles.

But in infants, the scalp, face, palms and soles are favored sites of scabies.

In people who are bedridden, the rash may occur only in sites that have been in constant contact with the sheets.

Once the mite lands on a new victim, it crawls from anywhere on the skin to its favorite sites at the rate of an inch a minute. The mite secretes a chemical to quickly digest the skin. By using its eight legs to move through the dissolved skin, a mite can burrow beneath the surface in less than three minutes.

Most of the itching is due to an allergic reaction to chemicals produced by the mite, because the rash and symptoms often are more widespread than the few sites where the mites are found.

The female mite deposits two or three eggs each day in the burrows. A few days later, the eggs hatch to produce larvae that mature to adults that wander to create new burrows or to be transferred to someone else to start a new infection.

The mites can survive about two days in bedding, clothing and house dust. In the body, the usual life span of an adult mite is a month, but some survive a few weeks longer.

Although there may be hundreds of mites, even millions in rare cases, a surprisingly small number can cause a tremendous amount of itching. In classic studies conducted among volunteers in England in World War II, the average number of mites found was 11 and often fewer than five.

The British studies suggested that scabies infections confer some degree of immunity. It was easier to cause a first infection among the volunteers than to reinfect people. But reinfections, once they take hold, develop more quickly, with fewer mites.

Scabies rarely kills and the mite does not spread other germs. But the scratching often leads to secondary infections caused by streptococcal and other bacteria, and these can produce life-threatening complications like kidney damage.

Most untreated scabies infections are chronic, giving rise to its nickname, the seven-year itch.

A standard treatment for scabies is lindane, an agricultural chemical once called gamma benzene hexachloride. It is usually applied once for up to 24 hours as a lotion and shampoo, and washed off. In adults the treatment is sometimes repeated.

Doctors sometimes treat people who have had contact with scabies, even if the mite is not found. To stop outbreaks in homes, schools, hospitals and nursing homes, doctors sometimes treat large groups of people and sterilize the bedding and underclothing.

Some cases of scabies seem to have become resistant to lindane, said Dr. Milton Orkin, an expert in scabies at the University of Minnesota. For this reason and because rare cases of convulsions from the treatment have been reported among children, even following a single application, scientists have been seeking newer therapies.

One is called permethrin or Elimite, and the synthetic is based on similar chemicals that were derived from the chrysanthemum. The first commercial production of insecticides made from the flower was in 1840, supposedly as the result of the observation of a woman of Dubrovnik, Yugoslavia. She picked flowers resembling daisies, discarded them in a corner when they withered, and later found them surrounded by dead insects.

The scabies mite was first described in Spain in the twelfth century, but doctors failed to link it with the disease until 1654. Medical historians say that the seventeenth-century finding gave scabies the distinction of being the first human disease with a known cause.

Throughout history scabies has been a scourge of wars. Scabies was notorious in the Napoleonic and American Civil wars, and it was the commonest skin disease reported among British soldiers in World War I.

In the late 1800s scabies was common in Austria, Scotland and Scandinavia, while it was rare in the United States.

The scabies mite causes epidemics that generally strike throughout the world and last about 15 years. Then a 15-year gap follows before the cycle begins again. The latest scabies epidemic began in the late 1960s. But, for unknown reasons, the current one has not abated as predicted.

One reason, some experts say, is that many cases are not typical and go undiagnosed, perpetuating the spread. Another is that doctors and people mistakenly believe scabies affects dirty people when, in fact, it can occur in affluent people who practice good hygiene.

The precise number of cases is not known because scabies is not a disease that doctors must report to health officials in the United States.

Some outbreaks have been linked to hospitals, while others have affected communities. In the 1970s an explosive outbreak struck one fifth of the town of Mexico, Maine, and forced closing of the schools.

Scabies most commonly affects children, but it also strikes those with weakened immune systems. In recent years scabies has been recognized increasingly among people with AIDS.

It is also common and can be hard to diagnose among the elderly, presumably because of waning strength of their immune defenses.

In 1983, Dr. Richard K. Scher, a dermatologist who practiced in Amityville, Long Island, and who now is on the staff of Columbia Presbyterian Medical Center in New York City, reported outbreaks of scabies at two nursing homes on Long Island that lasted for more than two years.

Credit for stopping the two outbreaks is given to a nurse who recognized what the doctors had overlooked: that the mites often hid under the victims' nails. Trimming the nursing home residents' nails and smearing them with lindane halted the outbreaks.

—LAWRENCE K. ALTMAN, M.D., October 1990

The Seven-Year Itch

The tormenting itch of scabies can puzzle diagnosticians, but treatment is relatively simple once the diagnosis is made.

Cause:

SARCOPTES SCABIEI, A MITE ONE SIXTIETH OF AN INCH LONG.

Incidence:

SCABIES IS NOT A REPORTABLE DISEASE IN THE UNITED STATES, SO THE PRECISE NUMBER OF CASES IS NOT KNOWN.

Symptoms:

EVEN A FEW MITES CAN CAUSE INTENSE ITCHING, ESPECIALLY AT NIGHT. THE MOST COMMON SITES IN ADULTS ARE SPACES BETWEEN FINGERS, WRISTS, ARMPITS, ELBOWS, BREASTS, GENITAL AREAS, NAVEL, BUTTOCKS AND TOES. IN INFANTS, COMMONS SITES ARE THE SCALP, FACE, PALMS AND SOLES.

Treatments:

LOTIONS AND CREAMS CONTAINING LINDANE ARE A STANDARD TREATMENT, BUT RESISTANCE HAS BEEN REPORTED. ONE ALTERNATIVE IS PERMETHRIN, SOLD AS ELIMITE, A SYNTHETIC BASED ON CHEMICALS DERIVED FROM THE CHRYSANTHEMUM.

Fearsome Fire Ants Stage
Relentless Campaign in South

IN THE ANNALS of entomological villainy, few insects are as despised, as feared and as meticulously investigated as the tiny imported fire ant, which attacks humans, animals, plants, other insects and even electrical devices.

Government and private researchers have spent hundreds of millions of dollars in unsuccessful efforts to eradicate the ferocious little arthropod, which is no larger than the "G" at the start of this sentence. Scientists are mapping its mitochondrial DNA and studying its most intimate chemical scents. Frustrated homeowners spend tens of millions more on the counterattack, with tactics ranging from pesticides to electric prods, even fighting fire ants with fire.

Yet true to its name, *Solenopsis invicta,* from the Latin for "invincible," has marched through the South like General Sherman. From its initial pincer-hold in Mobile, Alabama, where it arrived on a ship from South America half a century ago, invicta now infests 250 million acres in 11 Southern states.

Researchers say it has just established a successful colony in Santa Barbara, California, its first foray into the country's richest agricultural state. Entomologists say invicta also seems to be undergoing a fundamental evolution in its social system, establishing giant colonies with multiple queens instead of the smaller and more manageable single-queen colonies that had been characteristic.

"It's just a matter of time before we have fire ant infestations in California and Arizona that we cannot control," said Dr. William P. MacKay, a specialist in ant ecology and systematics at the Fire Ant Laboratory at Texas A & M University. "It's going to go all the way up the West Coast, all the way to Washington State." Colonies have been found as far north as Virginia.

The ant's inexorable spread has occurred not only in spite of, but partly because of, nearly four decades of determined campaigns to eradicate it with chemical pesticides. Through it all, the fire ant has thrived because the pesticides destroyed its enemies and paved the way for far larger colonies than had previously existed. An airborne pesticide assault in the 1960s and '70s, designed to eradicate the ant from American soil, has been described by the Harvard biologist Edward O. Wilson as "the Vietnam of entomology." A more modest attempt to control the ant with pesticide applications on the ground also failed.

The pesticides not only killed the ants' competitors, they also killed birds and other nontarget species. Mirex, in particular, left significant residues in human and animal tissues. Mirex was determined to be a possible human carcinogen in 1977, and the Government's aerial spraying was halted in 1978.

Today, in a reversal, most state and federal agencies now discourage the use of pesticides against the ant. They urge the public to learn to live with the ant, which is more a nuisance than a health hazard, and they hold out hope that scientists will enlist beneficial organisms to attack the ants or develop new compounds to disrupt their reproduction.

But agriculture officials said there was again growing pressure from farmers and ranchers to explore the option of spraying. The cost of such a new program would be prohibitive, the officials said.

Adversity seems to have brought invicta new strength. Initially the species had a single queen ant in each colony, and each colony was fiercely territorial. As a result, researchers said, the average density of infestation was rarely greater than 20 or 30 mounds an acre.

But researchers have recently found that single-queen colonies are being driven out by "supercolonies" containing multiple queens and tens of millions of workers. Researchers do not know what is causing the change, but they note that the older single-queen colonies, which were a nuisance, are now viewed as the best defense against the newer, more powerful colonies, which are an economic disaster.

In Texas, where the multiple-queen colonies now outnumber single-queen colonies, densities of 450 mounds an acre have been reported. In a draft report prepared for the Texas Fire Ant Advisory Board, the state Department of Agriculture estimated that fire ants caused $47 million in crop losses and pest control expenses in the state last year.

No national loss figures are available. "We don't see any hope at this point of eliminating the pest from the United States," said William A. Banks, a fire ant researcher at the United States Department of Agriculture's fire ant research service in Gainesville, Florida. "The hope is that we can find better ways of coping, so that we can learn to live with it."

The prospect of peaceful coexistence with invicta is hard for many Southerners to accept, especially for those who know its sting, which is often mistakenly referred to as a bite. The ant does bite, but only to anchor itself while it stings. Unlike bees, which eviscerate themselves with each sting, the fire ant stings repeatedly with a gusto that belies its size.

The bite is sometimes felt as a minor irritation, but the sting is immediately sensed as a burning sensation, worse than a mosquito bite but not as painful as a bee sting. It is common for people to be stung many times, since the ants are highly aggressive and swarm onto an intruder in seconds. Some victims have reported suffering hundreds of stings.

A sting later results in redness and swelling and, because invicta injects bacteria along with its alkaloid venom, creates a small pustule on the skin. The pustule, typically the size of a pinhead, later causes itching, and sometimes the pustules leave small brown scars that last for months.

Fire ant stings are potentially fatal to people who are allergic to bee stings and other venoms. Ryan Wingard, a 2-year-old in Anderson, Texas, nearly died one May when he was stung by fire ants in his yard. His mother noticed that he was having trouble breathing, and soon Ryan's lips turned blue, his tongue swelled and red blotches appeared on his body. He was rushed to the hospital and treated for a severe allergic reaction. For Ryan and many others in the South, playing or picnicking in the grass is now just a memory.

"I went to Willie Nelson's Fourth of July picnic and watched people popping up like popcorn, slapping their legs," said Dr. Edward Vargo, an entomologist studying fire ants at the University of Texas's Brackenridge Field Laboratory in Austin.

Dr. Vargo noted that the multiple-queen colonies can be elaborate structures that comprise hundreds of mounds covering dozens of acres, allowing worker ants—sterile females—to forage with astonishing efficiency. "Basically, anything that stands still for longer than 15 or 20 seconds is fire ant food," he said.

Food, to a fire ant, is a broad term. Its diet includes everything from insects to germinating crop seeds to the rubber expansion joints on highway bridges. The ants feed day or night, as long as the temperature is 70 to 95 degrees Fahrenheit.

The impact on the ecology is especially worrisome to researchers, who note that invicta drives out other ant and insect species, including native fire ants, which are less aggressive. Increasingly, landowners say the ants are driving off ground-nesting birds, lizards, rabbits and larger wildlife. The Small Animal Clinic in the College of Veterinary Medicine at Texas A & M has treated 28 fawns this year for severe fire ant stings.

Because invicta is omnivorous, it also has some benefits. Invicta attacks corn worms, boll weevils and other costly agricultural pests, and is actually welcomed on cotton and sugar cane fields. The fire ant has also shown a fondness for fleas. However, "Using fire ants to control the flea population does seem a lot like trading a headache for an upset stomach," said Kathy Palma, a researcher at Texas A & M. Fire ants do not live by food alone, Dr. MacKay at Texas A & M reported. He has found that the ant is attracted to electrical fields, the stronger the better. It is common for the ants to swarm on electrical relay switches inside air-conditioning units, until their charred bodies cause a short circuit. Municipal officials throughout Texas say fire ants are the leading cause of traffic light failures. Telephone companies say the ants can knock out switching devices. Dr. MacKay said the ants have been found in aircraft altimeters, in computers and even in the electronic gear in boats and submarines.

The ants are aquatic at times, stinging people in swimming pools and rivers. In flood conditions hundreds or thousands of workers will form a living raft to carry the queen and her larvae to safety. Some researchers suggest that such living rafts will help carry invicta west.

More commonly, the ants hitchhike in shipments of grass sod, trees and other nursery stock. They also spread as the queens seek new homes after their mysterious nuptial flights. Only queens and males develop wings, and because fire ants do not mate in captivity, and because it is difficult to observe tiny insects frolicking 500 to 800 feet in the air, researchers surmise that the mating is airborne.

On the first sunny day after rain, Southerners can see hundreds of winged male fire ants emerge from the soil and ascend on a sexual kamikaze mission. They fly only once, rising to several hundred feet, where they compete to mate with a flying queen who has summoned them by some unknown chemical signal. Only one will succeed, and then all the males, perhaps exhausted, fall back to Earth to their deaths.

"If they mate they die happy, and if they don't, they die anyway," said Dr. Daniel Clair, an urban pest management specialist with the Texas Department of Agriculture.

The fertilized queen sheds her wings, burrows and either starts a new colony or joins an existing one. She stores the sperm from her nuptial flight, and for the next five or six years produces offspring, often laying her own weight in eggs each day.

Chastened by the failure of pesticides, researchers are focusing on two areas: chemical growth regulators and biological controls.

Dr. Vargo and others are trying to isolate and synthesize the chemical signals, or pheromones, given off by invicta queens. Pheromones are chemicals that act on other individuals, unlike hormones, which act within an individual.

Among the queen pheromones of particular interest to researchers are signals that cause workers to assassinate rival queens and signals that keep immature females from developing sexually. If such pheromones can be developed, they could disrupt communications in the colony and the ants would eventually die.

The most promising chemical control, said Ronald Mulder of the Texas Department of Agriculture, is fenoxycarb ethyl, essentially a "birth control" agent that destroys the ovaries of sexual female ants. It is sold commercially under the brand name Logic and is believed to be much more environmentally safe than pesticides. To the frustration of homeowners, Logic is also slower acting, taking three to six months after a springtime application to take effect.

Dr. MacKay said there was also promise in biological controls. He and other entomologists who have visited invicta's native home in South America found that the ant was much less of a problem there than in the United States.

He said invicta mounds in Brazil are shared by several types of parasitic ants, beetles, wasps and flies. If one of these or some other natural fire ant enemy can be identified as a control agent and brought to the United States, assuming it does not pose a threat worse than fire ants, it could help keep invicta in check.

How about South American anteaters? "The ones that could do the job are four feet tall with claws that could take your arm off," said Mr. Banks of the United States Agriculture Department. "If we imported enough anteaters to have an impact, in a few years we'd be looking for something to control the anteaters."

—PETER H. LEWIS, July 1990

Power of Natural Pest-Killer Wanes from Overuse

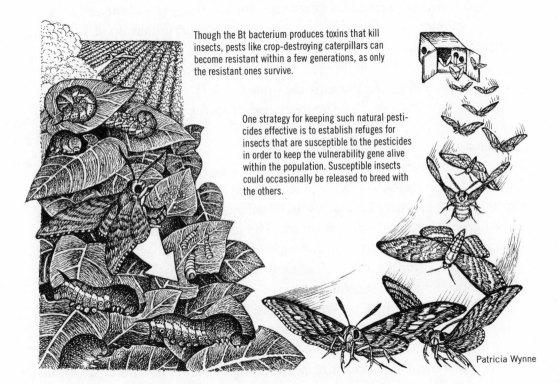

Though the Bt bacterium produces toxins that kill insects, pests like crop-destroying caterpillars can become resistant within a few generations, as only the resistant ones survive.

One strategy for keeping such natural pesticides effective is to establish refuges for insects that are susceptible to the pesticides in order to keep the vulnerability gene alive within the population. Susceptible insects could occasionally be released to breed with the others.

Patricia Wynne

WITH SURPRISING SPEED, insect pests are becoming resistant to natural toxins that scientists hoped would usher in a new era of biological controls and ring down the curtain on synthetic insecticides.

The natural toxins, produced by a bacterium called *Bacillus thuringiensis*, or Bt, are widely used in sprays to kill agricultural pests and forest scourges like the gypsy moth and spruce budworm. Bt toxin has been hailed as a perfect pesticide because it targets only certain caterpillars without harming either the insects' predators or leaving a poisonous residue on crops or trees.

The discovery that insects can develop resistance to Bt toxins in as little as two or three insect generations has cast a cloud over efforts to develop a wider range of biological pest controls. It has also raised questions about an important and imaginative project—that of engineering resistance into crop plants by endowing them with the genes to make their own toxins, including those produced by Bt.

Cotton plants and potatoes like this have already been developed and await government approval. Since more than 500 species of insect pests have become resistant to chemical pesticides, the effort to expand the use of natural pesticides is a matter of urgency, yet could be undermined before it gains much headway if resistance continues to emerge.

To preserve the usefulness of the Bt family, biologists suggest a range of tactics, including the paradoxical notion that farmers should keep a part of their crops untreated so the insect pest can thrive there. The idea is to keep susceptible individuals alive and prevent the population from being dominated by resistant insects.

The first cases of resistance to Bt were described in the journal *Science* by Dr. William H. McGaughey of the United States Department of Agriculture grain marketing laboratory in Manhattan, Kansas, and Dr. Mark E. Whalon of Michigan State University. The Indian meal moth, the Colorado potato beetle, the tobacco budworm and the diamondback moth have all developed resistance to Bt toxins, they wrote, and signs of weak resistance have appeared in two mosquito species.

Dr. Lester E. Ehler, an entomologist at the University of California at Davis, said the emergence of resistance after excessive and heavy-handed use of Bt was "quite predictable." Indiscriminate use of any pesticide is known to foment resistance since, under the pressures of natural selection, susceptible insects will rapidly die and any resistant forms that have emerged will proliferate in their place.

Dr. McGaughey has found that the Indian meal moth can develop resistance in only two or three generations, and other researchers report that the tobacco budworm can produce it in about a dozen generations.

Bt "has the potential for replacing a lot of chemicals that society wants to see taken off the market," Dr. McGaughey said, but "if we see rapid development of resistance, we may lose Bt and not have many safe materials left."

Historically, farmers have tended to take the easy route by applying synthetic pesticides in large doses.

"This same pesticide mentality is at it again" in the case of Bt, Dr. Ehler said. The sensible alternative, he said, is to "try to intervene intelligently as opposed to bringing out the heavy artillery," adding, "There is no magic bullet."

The approach he and other experts favor is to use Bt sparingly and in combination with other biological agents. "It's really important never to put all your eggs in one basket and to use a diversity of approaches," said Dr. R. James Cook of the Department of Agriculture's biological control research unit at Washington State University.

Dr. McGaughey and Dr. Whalon have suggested several tactics for preserving the effectiveness of Bt.

One tactic is to alternate one kind of Bt toxin with another, or with other insecticides. The assumption is that the various toxins will kill different genetic groups within the pest species.

For example, Bt has become the pesticide of choice for spraying Northeastern forests ravaged by gypsy moths. Dr. Ann Hajek and her colleagues at the Boyce Thompson Institute for Plant Research at Cornell University have found that a fungus, *Entomophaga maimaiga,* is also an effective killer of gypsy moth caterpillars.

Both Dr. Hajek and Dr. McGaughey say the fungus, if its use in practice proves out, might be alternated with Bt without the risk of creating a resistant population of gypsy moths. There are also hopes of adding a virus being developed by the National Forest Service to the anti–gypsy moth lineup.

In a related tactic, two or more seed lines of plants could be genetically engineered to produce different toxins, and the seeds of the different lines mixed in planting.

The risk of these approaches, however, is that if the second agent should induce resistance in the same group of individuals as the first, resistance to the first agent would develop much quicker.

Also, some Bt toxins are known to trigger resistance not only to themselves, but to related toxins as well. "We know about twenty of the Bt toxins," Dr. McGaughey said. "That's not an unlimited number, and if you have the misfortune to use one and cause resistance that elicits cross-resistance to fifteen of the others, you don't have much left to work with."

A third tactic is to apply a toxin in doses high enough to keep pests under moderate control but low enough so that some insects genetically susceptible to the toxin can survive and forestall resistant insects from dominating the population.

In a fourth tactic, plants would be engineered so the Bt toxin was expressed only in the part of the plant attacked by insects, like the fruit. The crop would survive, along with Bt-susceptible pests.

And in what Dr. McGaughey and Dr. Whalon say may be the best tactic of all, resistant individuals could be blocked from taking over the gene pool by simply establishing refuges, or preserves, for susceptible individuals, much as other wildlife is preserved. These refuges, in effect, would be bases from which susceptibility genes would be continually spread back into the general pest population, a notion supported by tests in a computer simulation.

Reserves could be created by mixing genetically engineered with ordinary plants, or by leaving whole fields untreated with Bt spray.

The ideal, Dr. Ehler says, would be to "look at the total ecology of a pest problem" and attack it with a combination of measured and controlled actions.

"For a given target pest," he said, "I would like to see two or three tactics in our arsenal that are compatible" so that it is not necessary to rely only on one. Otherwise, he said, the pesticide will soon outrun its usefulness.

In that case, he said, "You're on the pesticide treadmill; you're constantly adjusting to meet the resistance. You have to ask if that's the way you want to operate in the long run."

—WILLIAM K. STEVENS, December 1992

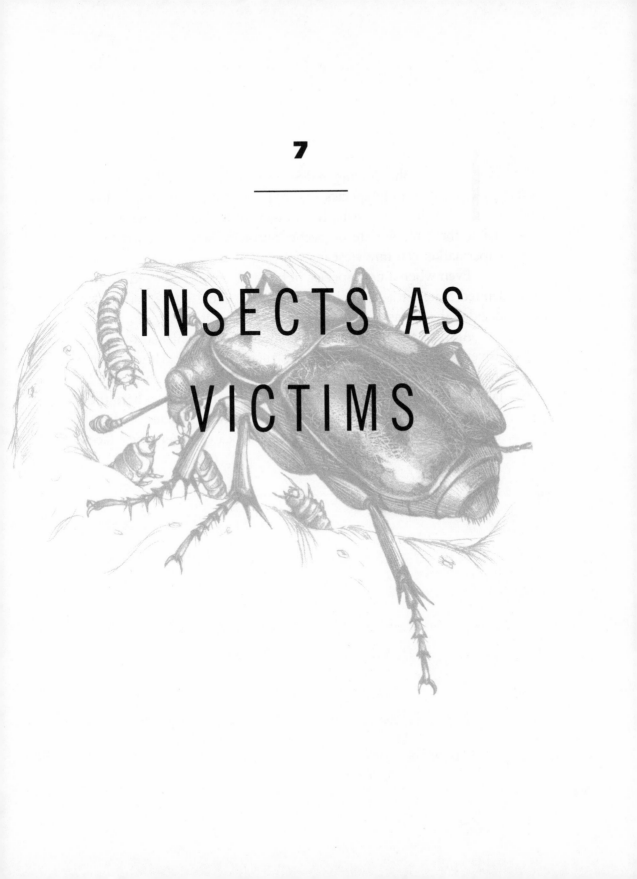

7

INSECTS AS VICTIMS

In theory the Endangered Species Act protects all species; in practice, only species that appeal strongly to people, like large, furry animals, have much of a chance of wriggling inside the tent. A mite or beetle whose habitat is needed by a supermarket can rarely escape oblivion.

Even when a new species of butterfly was discovered in the United States, the *Uncompaghre fritillary,* some conservationists decided there was not enough money to save it from extinction.

Abroad, thousands of insect species have vanished or been driven toward extinction by the destruction of tropical forests. For the showier species of butterfly, like the homerus swallowtail of Jamaica, there is hope of survival in butterfly farms, where the insects are raised for sale to collectors. But most species of insect arouse only aversion, and find no advocates when they face extinction.

Butterfly Farms May Be Last Hope of Rare Hunted Species

WITH A WINGSPAN nearly half a foot across, banded in black and yellow with iridescent blue patches, the homerus swallowtail is one of the most spectacular butterflies on earth. But its beauty and rarity have brought the gorgeous insect to the brink of extinction as poachers and collectors prowl its last refuges in the remote mountain rain forests of Jamaica.

A team of conservationists working with the Nature Conservancy and the Xerces Society, a group that protects endangered insects and their habitats, is trying to rescue the last homerus colonies. Their plan is to discourage the illegal poaching of the wild butterfly by local people and instead to train them how to cultivate the insect themselves in butterfly farms.

The most effective way to save rain forests is to provide local inhabitants with incentives to save them, not cut them down. The homerus project is part of a growing effort by conservationists around the world to save the habitats of endangered species by setting up cottage industries like butterfly farming. Projects for raising exotic butterflies have already begun in Ecuador and China.

The new farms are modeled after the successful example of Papua New Guinea. This tropical island is home to butterflies known as birdwings, so called because they can be mistaken for birds. Since the 1970s, the farms in New Guinea have been helping to turn an epidemic of poaching and destruction into a profitable industry.

Dr. Thomas C. Emmel, a professor of zoology at the University of Florida and a researcher on the homerus project, said the poaching problems that threaten homerus did not originate locally.

"The dealers are from the U.S. and Germany, the two chief offending countries," he said. "Most of the customers are in Japan. The dealers go to

Jamaica and they go to one of the towns near a classic collecting site and they put out the word that they are ready to buy. Maybe they bring in a few nets, give out their address and offer fifty dollars a piece for a butterfly that they can sell later for fifteen hundred. The average annual wage of the people is three hundred dollars. So a subsistence farmer says, 'The heck with farming, I'm going to go collect these butterflies.'"

Many butterflies are wasted because dealers will only pay the full price of $50 for specimens in perfect condition. Homerus's soft fragile wing edges appear to get damaged in the first two or three hours of flight through the forest, making most specimens caught in the wild much less valuable than those raised and carefully killed at a butterfly farm.

In theory, butterfly farming is simple. Just plant the water mahoe, the foodplant of the butterfly, near homes, on the outskirts of villages, in gardens and around the edges of the forest. Then let the butterflies do all the work. One homerus female will drop down out of her flyway high in the forest canopy to lay her eggs. The caterpillar will hatch, eat until it is nearly the size of a full-grown mouse, and magically metamorphose into an adult homerus.

Butterfly farming has its complications, however, and these remain to be worked out in the case of homerus. Most butterflies are extremely fussy about where and when they lay their eggs, and their caterpillar offspring can be quite demanding about what they eat and where they live. Homerus is not likely to be any different. Before any butterfly farms are in place, researchers will have to find out exactly what is necessary to bring the females in and keep their caterpillars alive.

Scientists are fairly sure that one problem homerus farmers will have to solve will be how to keep their farms as wet as the wettest rain forests. "The favorite habitat for homerus and its foodplant, whose name means the water-filled tree, is along streams" in water-saturated areas that get 300 to 400 inches of rainfall a year, Dr. Emmel said.

In order to learn more about the secret habits of this rare butterfly, scientists will be enlisting the help of the experienced butterfly farmers at Butterfly World, a tourist attraction and research facility near Fort Lauderdale, Florida. Ron Boender, the center's founder, said it had enclosed flight areas full of tropical plants, complete with waterfalls and mist systems where the initial experiments could easily be done.

It might seem surprising that avid collectors, presumed butterfly lovers, would go so far as to help drive a species to extinction and be willing to break international law to do it. But for some the desire to own the rare specimen is irresistible. "It's like buying a stolen Rembrandt," said Mr. Boender. "You can't display it. You can't say you have it. I guess it's like the drug trade—you're hooked."

In Papua New Guinea, the Coast Guard patrols the northern shores of the island nation to keep out people seeking to enter illegally to poach birdwings. "The Papuan Government has intercepted these boats," said Dr. Emmel. "They've arrested these people, thrown them in jail, fined them heavily. Sometimes they stay in jail for years. And yet they keep coming back." Because some birdwings can bring $7,500 a pair on the black market, the incentive to take the last of what is left remains strong.

As the number of rare species grows, the lure of the profits of rarity, or extinction, has increased. Buyers present a potentially devastating threat to the endangered species that they covet, and new reports of illegal collection of rare butterflies are turning up all over the world.

While he says that a butterfly farm could be crucial to the survival of homerus, Michael Parsons, a founder of the New Guinea farm project, warned that "to say that butterfly farms are the be-all and end-all of saving butterflies is nonsense." One lesson from New Guinea, he said, was that no amount of money spent on butterfly farms could keep a species safe unless it was part of a larger government-sanctioned project to preserve the rain forest.

While aware of the need for a variety of coordinated conservation projects, leaders of the homerus project are simply trying to find enough money to get the butterfly farms going before the homerus becomes extinct.

The fate of homerus is very much in doubt, Dr. Emmel said. "It depends on the success of the fund-raising drive," he said. "It shouldn't cost more than twenty-five thousand dollars to set up. Think of the payoff. It's a modest investment."

—CAROL KAESUK YOON, April 1992

Trying Times for the American Honeybee

AFTER MORE THAN 350 YEARS of peaceful existence, the American honeybee is under assault.

Modern transportation has brought new enemies from across the oceans into the environment of the honeybee, including mites that are decimating domestic hives in many states and are threatening to spread nationwide. And ill-tempered descendants of African bees, released by mistake 30 years ago in Brazil, have completed their march to the Texas border and are ready for an invasion that threatens to displace the relatively mild-mannered domestic bee.

Bee experts say they hope to save American beekeeping with aggressive research, modern management and ancient techniques, like those of a 92-year-old English monk who laboriously breeds mite-resistant bees by hand. But the industry will be greatly changed.

"Twenty years ago, it was easy to raise bees," said Bob Brandi, a Los Banos, California, honey producer who maintains 4,000 colonies and is president of the American Beekeeping Federation. "It didn't take much to manage your colonies, the bees always seemed to be good and you could make a decent living. But in three years, if you are not a truly professional beekeeper who aggressively deals with parasites and other problems, you won't be around."

Many of today's bee problems stem from pests inadvertently imported from other areas of the world.

"What has happened to our bees?" asked Dr. Thomas E. Rinderer of the Agriculture Department's Agricultural Research Service in Baton Rouge, Louisiana. "Jet planes have happened."

The biggest current threat is the tracheal mite *Acarapis woodi,* an infestation that has swept American colonies since 1984. The mite, long established in Europe, resides in the trachea or breathing tube of the bee and restricts oxygen intake. American honeybees, all descendants of European species, have not had to deal with this pest for centuries and have therefore lost natural resistance to it, insect specialists said.

Tracheal mites weaken hives by sucking fluids from bees and reducing their capacity to make honey or pollinate plants. Beekeepers treat infected hives with menthol crystals, which produce fumes that clear the bees' windpipes of mites. But this treatment is only marginally effective because cold winter temperatures keep the menthol from vaporizing while unusual heat can cause too much evaporation, driving bees from their hives.

Dr. Rinderer of the Agricultural Research Service and his colleagues hope to stop the infestation with the help of British bees. The researcher acquired a set of new honeybee queens from Brother Adam, a monk from the Buckfastleigh Abbey in England, who has practiced beekeeping for more than 75 of his 92 years.

"The brother breeds bees that rarely have tracheal mites," said Dr. Rinderer. "We don't know exactly why they are resistant, whether it has to do with resistant genes or some other mechanism, but we want to see if they can serve as the basis of stopping the mites here."

Last week the 36 queens that descended from 14 queens that survived the trip from England were released from a six-month quarantine. The bees will be propagated further and their descendant queens will be at the center of 600 colonies that will be placed later this year in the north-central and south-central parts of the country to see if they can stay free of tracheal mites, Dr. Rinderer said.

The second mite problem plaguing honeybees involves one called *Varroa jacobsoni*, a larger, external parasite of adult bees and their developing larvae, or brood. First discovered in the United States in September 1987, the *Varroa* is a mite that normally affects Asian honeybees.

The *Varroa,* which experts said can wipe out a hive within three years, lays eggs on developing bee larvae and the young mites feed on developing bees, making them smaller and weaker than normal.

Adult mites feed on the fluids of adult bees and also make the bees more vulnerable to disease viruses. Bees normally are resistant to these viruses because they primarily contract them orally and the insect's gut stops the pathogens. But the feeding mites transfer the viruses directly into the bees' blood.

The *Varroa,* which has spread up the East Coast from Florida to Maine and into the Midwest, is vulnerable to one insecticide. But beekeepers, who say they have low profit margins, said it was very expensive to use and questioned its effects on the bees. In addition, scientists worry that the mites may develop resistance to the pesticide if it is widely used.

Dr. Roger A. Morse, an entomologist who studies bees at Cornell University, said the *Varroa* might prove to be the greatest threat, since it is difficult to diagnose early hive infections, and bees do not appear to have any natural resistance to the pest.

The importance of the $150-million-a-year beekeeping industry, which in good years produces 200 million pounds of honey, far exceeds its size. Owners move thousands of colonies around the country each year so bees can pollinate crops valued at up to $20 billion a year. Bee specialists said this extensive migration is a factor that complicates disease and parasite control.

As if the recent mite problems were not enough, the long-awaited arrival of the aggressive Africanized bees is at hand. What is believed to have been the first swarm of these to reach this country has crossed the Mexican border into the Rio Grande Valley of Texas. The bees have been moving northward at 200 miles a year.

Beekeepers and others fear that the Africanized bee, earlier nicknamed the "killer bee" because of the viciousness of its attacks, may displace its calmer cousin in the United States and devastate the industry. The Africanized bees, descendants of wild bees imported from Africa that escaped a breeding experiment in Brazil in 1957, are harder to handle and do not produce as much honey as domestic bees.

Dr. John G. Thomas, a bee expert at Texas A & M University, believes the industry can reduce the impact of the bees through new colony-management techniques, like wearing more protective clothing, using bigger smoke pots to calm the bees during honey harvesting and removing Africanized queens when they are found and replacing them with their European-derived counterparts.

Scientists have long hoped that as the Africanized bees mate with domestic honeybees, the hybrids would acquire the desirable traits of both, including what may be more disease resistance in the African species.

The Africanized bees at the front of their expansion are less aggressive than other Africanized bees, he said, perhaps because they have been mating with the milder, resident bees. But after they have been in an area five to seven years, they seem to get meaner, perhaps because the purer Africanized bees catch up with the leading edge of the expansion and start inbreeding again.

Dr. Rinderer of the Agriculture Department said he and several colleagues had new evidence that Africanized bees in Argentina and the Yucatán Peninsula of Mexico were being hybridized with European properties.

"The more Europeanized these bees, the better," he said. "It is conceivable that our troubles will not be as bad as predicted. This genetic diversity might produce quality bees that ultimately might be desirable."

—WARREN E. LEARY, January 1991

Rare Butterfly Consigned to Extinction

SCIENTISTS SAY the country's most recently discovered species of butterfly is about to become extinct. But rather than taking what they call "heroic measures" to save it, they are advocating a more hands-off policy that they predict will soon have biologists counting one less species among North America's butterflies.

Like concerned family members gathered around a loved one's deathbed, the biologists who studied this small alpine insect, the Uncompahgre fritillary butterfly, have begun the mourning process, already sadly shaking their heads. They say there is no clear way to save the butterfly, so rather than spending precious conservation dollars to try to resuscitate the species, they suggest letting it go.

"This is probably a fairly heretical stance to some people," said Dr. Hugh Britten, a conservation biologist at the Nevada Biodiversity Research Center in Reno who is an author of a paper about the species. "It's not very easy for me to say I am presiding over the extinction of this species. But in the world of science, so what? Your personal feelings are not relevant and they shouldn't be."

While Dr. Britten and his co-authors, Dr. Peter Brussard, president of the Society for Conservation Biology, and Dr. Dennis Murphy, a conservation biologist at Stanford University, recommend continued monitoring and guarding of the survivors, their ominously titled study, "The Pending Extinction of the Uncompahgre Fritillary Butterfly," makes clear their vision of the future.

"Remember that the whole funding for endangered species is one pie," Dr. Britten said. "We're saying there's not much we can do that's sure to work and we ought to leave those funds in there to be applied somewhere else where they can work."

Among conservationists, the recommendation on this butterfly, which measures barely an inch across its wings, has already elicited everything

234

from hearty approval to anger. Some say the researchers are making exactly the kind of hard decision they need to make and have been promising to make, forgoing expensive last-ditch efforts to conserve doomed single species.

"I think they're right on target with no heroic measures," said Dr. Paul R. Ehrlich, the Bing Professor of Population Studies and president of the Center for Conservation Biology at Stanford. "I can't get excited about captive breeding of an alpine arctic ecosystem species that may go extinct on its own. It'd be much better to work much harder to save the species of the Northwest—to save whole huge habitats."

But others warn that the recommendation may be typical of many examples to come in which biologists subscribe to a triage mentality, too quickly writing off one species in favor of others in the name of economic prudence.

"Personally, I find the triage idea ethically repugnant," said Dr. Reed Noss, editor of *Conservation Biology*, the journal in which the butterfly report was published. "We too readily assign species to the category of hopeless. It's too easy for political reasons to say we can't do anything about that species, that it would put too much of a burden on our economy. I probably wouldn't have concluded the same thing."

Dr. Jeffrey Glassberg, president of the North American Butterfly Association, which has its headquarters in Morristown, New Jersey, suggested that the constraints of the triage mentality may in some ways be self-imposed. "I think these are self-fulfilling prophecies: 'I have limited resources and we're not going to do anything.' I don't know that I'd want to give up so easily."

Named after the site in the San Juan Mountains of Colorado where the species was discovered in 1978, the Uncompahgre (pronounced un-kum-PAH-gray) fritillary—an unremarkable-looking black and brown mottled butterfly—has always been rare, often numbering in the hundreds at each of its two known locations. But more recently, the species seemed in precipitous decline, virtually disappearing from Uncompahgre Peak and declining at its other known site, Red Cloud Peak.

While acknowledging that the exact reasons for the populations' year-to-year ups and downs remain mysterious, the researchers blame the general decline of the species on grazing that once was allowed in the butterfly's alpine meadows, on overzealous collecting by those anxious to have this newest known and very rare species, and—most important—on a warming climate.

Thought to be a relictual arctic species left behind on Colorado's mountaintops when the last glaciers retreated some 10,000 years ago, the species, *Boloria acrocnema*, has been trapped in what these scientists say has been more than a decade of uncomfortably warming weather.

"As far as management implications go, there's essentially nothing we can do about it," said Dr. Britten, who is a member of a team appointed by the government to write a recovery plan for this endangered species. "When you turn the heat up, it causes problems and the butterfly has to move upslope, and there is no more upslope. It's being ecologically squeezed off the top."

All the experts seem to agree that if the species is naturally doomed, then the only responsible action is to let it go and allocate limited resources elsewhere. The problem comes in deciding when to say when.

Many expressed concern that the researchers may have been a bit too hasty in tolling the death knell for this species. Knowing the butterfly is indeed doomed, they say, is not as simple as it might appear to be. Researchers point out that just last month in Southern California, the Palos Verdes blue butterfly, long presumed extinct, turned up very much alive.

Amy Seidl, a graduate student at Colorado State University who is the only researcher now studying the Uncompahgre fritillary, said her recent work indicated that the situation was not as dismal as the researchers suggested. Since the downward trend observed until 1991, Ms. Seidl said she had witnessed a stabilization and even an increase in the butterfly's numbers.

The butterfly's population numbered nearly 2,000 last year at Uncompahgre Peak, according to Ms. Seidl's study, up from estimates of roughly zero in 1990 and 1991. She also estimated a population of 1,300 at Red Cloud Peak, up from 400 in 1990 and 400 to 1,000 in 1991. These are considerable increases for a butterfly thought to be on its way out.

Given the butterfly's apparent comeback, some researchers suggested that attempts at the heroic measures of captive breeding and introduction to additional mountain meadows were in order.

"Nobody can see into the future that far," Dr. Larry Gall, a curatorial affiliate at the Peabody Museum at Yale University, said of the predictions of pending extinction. "I think it would be useful for them to attempt an introduction."

Dr. Gall and Dr. Felix Sperling of the University of Ottawa discovered the Uncompahgre fritillary, the only new butterfly species found in the United States in the last three decades.

Dr. Britten and his colleagues, however, said they were not hopeful about captive breeding or introductions of wild butterflies to new areas, citing poor success rates in the past.

Dr. Gall responded, "Sometimes it's awful but sometimes it's good. It's not unreasonable to try."

Despite arguments against the costliness of such ventures, Terry Ireland, a Fish and Wildlife Department biologist who is a member of the butterfly recovery team, said that for this species, even the most heroic of measures—captive breeding—is estimated to cost no more than $5,000 a year. According to a Fish and Wildlife Survey, in 1991 federal and state agencies spent an estimated $177 million on the recovery of federally listed endangered species alone.

But so far, the recovery plan closely follows the hands-off recommendations of the researchers, a proposal of simple monitoring of the population, modest attempts to search for new populations and studies to learn more about the biology of the insect.

Even if funds were available and researchers willing, heroic measures can have their risks. Dr. Britten, well aware of Ms. Seidl's findings of increased numbers, remains wary of disrupting the butterflies with removals for breeding or introduction elsewhere. He cautioned that insect populations often fluctuate greatly, with apparently healthy populations quickly dropping back to low numbers for no apparent reason. The problem is that when butterfly numbers drop down now and again, bad weather or other harsh conditions can keep the vulnerable populations from bouncing back.

While the more pessimistic researchers argue that such fluctuations make it easy for the Uncompahgre fritillary to be permanently knocked out, others counter that it also makes it more difficult to predict extinction.

Adding more uncertainty to the status of the butterfly is the specter of yet undiscovered colonies. Ms. Seidl is optimistic that there are still unknown populations of the butterfly. As part of her continuing study, she

to scale the wet, northeast slopes of the San Juans, the sort the butterfly would most likely favor.

Even Dr. Britten, who has made numerous searches, admits it is impossible to be sure whether or not there are hidden colonies in the wilderness
of the San Juans. It is particularly difficult because the butterflies are visible
and in flight for only about three weeks a year, in July.

"All this means we can't be very confident that we've looked everywhere at the right times," he said. "There have been reports of additional
colonies by one other lepidopterist, who is refusing to reveal where they are.
You get reports like that and you think, maybe this person's right."

Amid the difficulty of assessing the health status of an endangered
species, conservationists continue to struggle with making what may be life-
or-death decisions for the animals and plants they study.

Ann Swengel, international co-editor for the annual July 4 butterfly
count run by the Xerces Society and the North American Butterfly Association, said, "When is it okay not to try to save a species? I think conservationists are groping with this now.

"In the end, it's an opinion, a judgment call just like everything in conservation. You never know what would have happened if you'd done something else."

—Carol Kaesuk Yoon, April 1994

Captive Breeding Project Offers New Hope for Beleaguered Beetle

IN A FORMER RESTROOM at the Roger Williams Park Zoo, a dozen or so American burying beetles have just emerged from underground—well, from the depths of a five-gallon flowerpot—and are happily munching mealworms in one of the more unusual captive breeding programs in the 23-year history of the Federal Endangered Species Act.

The beetle, named for the adults' practice of interring the carcass of a small mammal or bird as a ready food supply for the larval young, is among the world's rarest insects. Nearby Block Island is home to the last known natural population east of the Mississippi River. But with a little bit of luck and a small amount of cash, the zoo could have a few hundred American burying beetles (*Nicrophorus americanus*) ready for release next summer at a Massachusetts Audubon Society sanctuary on Nantucket Island. The United States Fish and Wildlife Service hopes to reestablish the species there on a small piece of its historic range. The insect was last found on Nantucket in 1926.

Perched on the edge of a beat-up sink left over from the cramped facility's previous duty, Dave Wetzel, the zoo's general curator, described the captive-breeding project as "a mixed success" so far. While the zoo has provided more than 140 adult burying beetles for release on Nantucket, most of the larvae the zoo raised earlier this year failed to emerge from the pupal stage. "They simply disappeared," he said.

Mr. Wetzel gently retrieved one of the beetles from its hiding place, in the folds of a paper towel lining a clear plastic box, and held the animal up for inspection. The inch-and-a-half-long insect is glassy black with brilliant flame-orange markings on its wing covers (or elytra) and a large protective plate (or pronotum) behind its head.

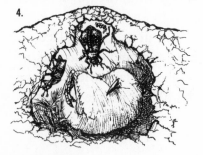

Buried Larder for the Family

Endangered American burying beetles (*Nicrophorus americanus*) shown greeting emerging larvae above have been released in a pilot program on Penikese Island, off Cape Cod. They provide for their offspring by storing corpses. At left, they work on a herring gull chick.

1. The male and the female get together at the site of a fresh corpse. Before mating they prepare the chick.

2. The corpse has been rotated and is being buried headfirst. In some cases the feet are left protruding out of the ground.

3. The beetles further compact the fully buried corpse into a tight capsule in the ground. The beetles apply secretions that control the rate of decay.

4. One of the beetles strips the corpse of its down and stows the feathers in a corner. Both excavate a shallow cavity in the chick by feeding from the top; they regurgitate liquid nourishment into it, to be tapped by growing larvae.

240

Michael Rothman

Decades ago, insect collectors chased the American burying beetle as a prize for their boxes of pinned specimens, so conservationists know that the species was once widespread east of the Rocky Mountains. There are records from 35 states, which show that the insect was found as far north as the Upper Peninsula of Michigan and as far south as Corpus Christi, Texas.

The beetle's decline began a century ago; by the 1920s, it had vanished from the Northeast mainland. The last report from any site on the Atlantic Seaboard, other than Block Island, dates to 1947. When the species was added to the endangered list in 1989, Oklahoma had the only known population west of the Appalachians, although others have since been discovered in nearby states. The World Conservation Union, which puts out a "red data book" on threatened animals, says the beetle has experienced "one of the most disastrous declines of an insect's range ever recorded."

Twenty North American insects are listed on the federal endangered species list, but captive breeding has been attempted for only one other species: the Schaus swallowtail butterfly in Florida.

In a pilot program, 197 adult American burying beetles were released from 1990 to 1993 on uninhabited Penikese Island, once used for a leper colony, off Cape Cod, Massachusetts. Dr. Andrea Kozol of Sudbury, Massachusetts, an expert on the species who raised those insects in a laboratory at Boston University when she was a graduate student, said at a recent meeting of the Ecological Society of America that a self-sustaining population appeared to have been established on the island, now a state wildlife refuge.

Dr. Kozol said the American burying beetle could become a model for similar insect reintroductions "as less charismatic invertebrate animals receive some long-deserved attention." Burying beetles are an important arm of nature's cleanup crew, she said, and they have long been "a favorite subject of naturalists and behavioral ecologists because of the extensive parental care both males and females provide to offspring."

Michael Amaral, an endangered-species biologist at the Fish and Wildlife Service's New England field office in Concord, New Hampshire, said the rare beetle was "the perfect candidate to get past the icky-bug attitude most people have toward insects. " Mr. Amaral said he wanted to bring the species back in the East without the kind of controversy or expense that often accompanies endangered-species programs. "We call the American burying beetle 'the California condor of the insect world' because of its large size and carrion-eating habits," the biologist said. "But we don't have a con-dor-size budget." He estimated that less than $10,000 had been spent on protecting and restoring the beetle in the East.

The Providence zoo took over the beetle breeding project from Dr. Kozol in the fall of 1994. "Our start-up costs came to about $1,200," Mr. Wetzel said. "We had to buy an air-conditioner to keep the temperature cool year-round, a refrigerator to hold rats plus some pots from a garden shop. And we've put a lot more beetles back in the wild than they've managed with condors," he added with a smile.

Theories abound to account for the American burying beetle's disap-pearance: the loss or fragmentation of the insect's habitat; an increase in the number of scavenging mammals, like raccoons and skunks, that compete for carrion; the heavy use of pesticides, like DDT; or some exotic disease. However, 14 smaller species of burying beetles are widely found in North America, and Christopher Raithel, a biologist with the Rhode Island Divi-sion of Fish and Wildlife, said that no one had come up with a reason why one species would decline when all of its relatives remained relatively com-mon across their range.

American burying beetle populations, Mr. Raithel noted, "were largely gone 25 years before organochlorine pesticides were broadly applied." Mr. Amaral said the insect "confounds our notions about every species having a preferred habitat by appearing to be a habitat generalist," adding, "We've found it in maritime grasslands on Block Island and oak-hickory forests and pastures in Oklahoma."

But burying beetles need a reliable supply of dead animals in order to breed, and *Nicrophorus americanus* needs bigger carcasses than any of its kin. Deer mice and meadow voles are abundant but too small; a red squirrel, cot-ton rat or mourning dove is a perfect size. So, too, was the extinct passen-ger pigeon. Flights of millions, if not billions, of passenger pigeons darkened the skies in the mid-1800s, but the last wild one was shot around 1900, and some scientists wonder whether the loss of such a huge food base led to the American burying beetle's collapse.

Mr. Amaral said the insect might have survived on Block Island because of a large population of ring-necked pheasants there and the absence of scav-enging mammals. The pheasants lay a dozen or so eggs, but only three or four young survive. "The beetles make their living off the dead chicks," he said. On 70-acre Penikese Island, the beetles bury dead nestlings from a large colony of herring gulls.

The biologist added that the American burying beetle had disappeared from Long Island in New York after the use of whole fish, like shad and herring, to fertilize agricultural fields was banned in the 1920s.

Burying beetles are nocturnal, warm-weather fliers that use keen chemoreceptors in their antennae to find carrion. "There'll be a battle real over the carcass," Dr. Kozol said, "with boys fighting boys and girls fighting girls until one pair is left." The losers are driven off, and the winners dig a hole under the head of the dead animal, and, by dawn, they have interred the entire carcass. In the next 48 hours, they strip the body of feathers or fur, roll it into a baseball-size package and coat it with oral and anal secretions to retard decay.

All this activity forms a roughly circular subterranean chamber where the adults remain to care for their brood and protect the carrion from raids by other beetles. That kind of parental care is common only among social insects, like bees. The female beetle lays her eggs in an adjacent tunnel, and as many as 30 larvae hatch three days later.

"The larvae make their way to the carcass, and at first they beg for regurgitated food just like baby birds," Dr. Kozol said. "They rear up and stroke the parent's mandibles." The larvae will pupate in the soil near the brood chamber, emerge as adults in 48 to 60 days and overwinter until the next breeding season.

The Providence zookeepers skip the battling-beetle stage. They pair off adults raised from larvae collected on Block Island, fill a black plastic pot halfway with wet garden soil, compact the dirt by pounding it with a brick and provide the insects with a dead laboratory rat. "The males chase the females almost immediately in an attempt to copulate," Mr. Wetzel said. However, some captive pairs prepare carcasses but fail to produce larvae.

When adult beetles are released on Nantucket, they are marked for identification with a notch in the elytra, paired with a mate and buried atop a carcass. Biologists return to the island 10 days later to count the larvae at each site, and they monitor the adult population with pitfall traps—a one-quart glass jar buried to the rim and baited with ripe chicken meat.

Mr. Amaral said the year-to-year population of American burying beetles on Block Island was estimated at a few hundred and was considered stable.

But the biologist, who is the national recovery coordinator for the species, said he could not guess at the insect's numbers in the six states west of the Mississippi where it is now known to occur.

In 1989, scientists believed that a single Oklahoma county, 1,300 miles west of Block Island, held the world's only other population of American burying beetles. When the Fish and Wildlife Service blocked the construction of a road through the insect's habitat in 1991, the incident became a regional cause celebre. A front-page headline in The Tulsa World read, "Flesh-Eating Beetle Blocks Highway."

Since then, intensive searches have found the beetle in a dozen other Oklahoma locations as well as in Arkansas, Nebraska, South Dakota and, this summer, Kansas and Iowa. Some of these populations, Mr. Amaral said, are represented by only a single specimen. "We've determined that the beetle survives in more of its historic habitat on the western periphery," he said. "But the species is gone from 90 percent of its former range, and I would argue that its listing as endangered is still appropriate."

Mr. Amaral said, however, that it was no longer necessary for the Fish and Wildlife Service to take legal steps to protect the insect's habitat in the West. Plans are being made to move 50 pairs of American burying beetles from a military base in Oklahoma to Ohio, he said, in an effort to establish a mainland population east of the Mississippi.

—LES LINE, September 1996

8

THE LATEST FROM THE FIELD

Historians may argue over the importance of this president or that, but in the long march of civilization the events that made an indelible difference to human welfare are revolutions like the invention of agriculture ten thousand years ago, or the discovery of antibiotics.

It is truly humbling, therefore, to discover that both agriculture and antibiotics were first invented 50 million years ago by another social animal, the ant. A particular family, the attine or leaf-cutting ants, learned how to cultivate a mushroom-like kind of fungus which they farm in underground chambers and feed with mulched leaf fragments.

The fungus, however, as described in the first article of this chapter, has a devastating enemy, a mold called *Escovopsis* which exists nowhere but in ants' nests. *Escovopsis* can run riot through a colony's fungus gardens unless kept severely in check. And the ants, amazingly, have developed a way of doing so: with the antibiotics produced by a *Streptomyces* bacterium.

A second article describes another intimate association between *Heliconius* butterflies and the passion-flowers that are their foodplant. Both the flowers and the butterflies stand out for their beauty. But they are intertwined by sharing a dark secret: both are poisoners that can afford to flaunt their elegant colors because of their chemical defenses.

The passion-flowers are laden with cyanide to protect against leaf-munching creatures like caterpillars. One predator, the *Heliconius* butterfly, has broken the code and learned how to detoxify cyanide. It has also learned how to make its own cyanide-producing chemicals. The tussle between the butterfly and the passion-flower seems to have driven both to radiate into many species.

Geneticists are slowly beginning to unravel the biological programming that drives the insect world. A third article describes a remarkable gene that first came to light in the laboratory fruitfly. Fruitflies with one form of the gene are explorers who roam far and wide in search of food; flies with the other form stick around the same old pile of rotting fruit. The rover/sitter gene also exists in honeybees but they use it in a different way. The gene has only one form which is switched off in young bees who stick around the hive doing domestic duties. As the bees mature, the gene is switched on, and its protein product somehow rewires the bee's brain to turn it into a forager.

Lastly, a grisly tale of a parasitic wasp that controls its host's behavior. The wasp larva preys on an orb spider and has learned how to force the spider to build a special web to the larva's liking, presumably by secreting some chemical that affects the spider's nervous system.

The research described in these four articles shows the deep insights that researchers are now gaining into the insect world, as well as the complexity and sophistication of these ancient creatures and their remarkable strategies for survival.

———————

For Leaf-Cutter Ants, Farm Life Isn't So Simple

LEAF-CUTTING ANTS and their fungus farms are a marvel of nature and perhaps the best known example of symbiosis, the mutual dependence of two species. But the textbook accounts, it turns out, do not tell even half the story. From research in the past five years the ants' symbiosis has emerged as far more intricate than it appears, involving not two but at least four species, their lives knotted together in a ruthless yet highly successful struggle for survival.

The ants and their agriculture have been extensively studied over the years, but the recent research has uncovered intriguing new findings about the fungus they cultivate, how they domesticated it, and how they cultivate it and preserve it from pathogens.

For example, the fungus farms, which the ants were thought to keep free of pathogens, turn out to be vulnerable to a devastating mold, found nowhere else but in ants' nests. To keep the mold in check, the ants long ago made a discovery that would do credit to any pharmaceutical laboratory.

The fact that there was still so much to learn about leaf-cutter ants and their agriculture, biologists say, only underlines how much remains to be discovered about the world's plants and animals.

Ants invented agriculture 50 million years before people did, and the leaf-cutters, members of a large family called the attine ants, practice the most sophisticated example of it. They grow their fungus, a kind of mushroom, in underground chambers that can reach the size of a football. A single leaf-cutter nest may contain a thousand such chambers, embedded in an underground metropolis up to 18 feet deep, and support a society of more than a million ants.

These ant communities are the dominant plant-eaters of the Neotropics, the region comprising South and Central America, Mexico and the Caribbean. Biologists believe some 15 percent of the leaf production of tropical forests disappears down the nests of leaf-cutter ants. In the nest the leaves are shredded and inoculated with the fungus, which digests them and is in turn eaten by the ants.

The ants' achievement is remarkable—the biologist Edward O. Wilson has called it "one of the major breakthroughs in animal evolution"—because it allows them to eat, courtesy of their mushroom's digestive powers, the otherwise poisoned harvest of tropical forests whose leaves are laden with terpenoids, alkaloids and other chemicals designed to sicken browsers.

So precious is their particular strain of fungus that the ants' virgin queens, before their nuptial flight, secrete a mouthful with which to seed the garden of their new nest. The worker castes they produce are so tailored to the craft of fungus gardening that they come in made-to-fit sizes—large ants to saw off leaves, medium ones to shred them and miniature workers to seed them with fungus and clean off all alien growths.

Fungus growing seems to have originated only once in evolution, because all gardening ants belong to a single tribe, the descendants of the first fungus farmer. There are more than 200 known species of the attine ant tribe, divided into 12 groups, or genera. The leaf-cutters use fresh vegetation; the other groups, known as the lower attines because their nests are smaller and their techniques more primitive, feed their gardens with detritus like dead leaves, insects and feces.

A question that has long perplexed ant biologists is whether the funguses cultivated by attine ants are all descended from a single ancestor, just as the ants are. The issue was hard to settle because the ants' gardening habits prevent the fungus from forming mushrooms, the spore-bearing stage by which mycologists tell one fungus from another. In 1994 a team of four biologists, Ulrich G. Mueller and Ted R. Schultz from Cornell University and Ignacio H. Chapela and Stephen A. Rehner from the United States Department of Agriculture, analyzed the DNA of ant funguses. The common assumption that the funguses are all derived from a single strain, they found, was only half true.

The leaf-cutters' fungus was indeed descended from a single strain, propagated clonally, or just by budding, for at least 23 million years. But the lower attine ants used different varieties of the fungus, and in one case a quite separate species, the four biologists discovered.

After further study, three of the biologists, Dr. Mueller, Dr. Rehner and Dr. Schultz, reported last year that funguses grown by lower attine ants fell into four groups of varieties, as if the ants had domesticated wild funguses at least four times in evolutionary history. Two of those occasions must have been quite recent because the biologists identified free-living counterparts for two of the four fungus groups they found in the ants' gardens. A single variety of fungus is grown in each nest, but most lower attine species cultivate at least two of the four fungal lineages, suggesting that varieties are exchanged among species every so often, the biologists concluded.

What evolutionary force could be driving these two patterns of fungus gardening, the pure clone cultivation of the leaf-cutters and fungus exchange program of the lower attines? The answer, or part of it, has been divined by Cameron R. Currie, a Ph.D. student in a climate no wild attine ever reached, the University of Toronto. Trained as an entomologist, Mr. Currie was attracted to the ants because of his interest in symbiosis and in the cheaters who take advantage of that mutualism.

The pure strain of fungus grown by the leaf-cutters, it seemed to him, resembled the monocultures of various human crops, that are very productive for a while and then succumb to some disastrous pathogen, such as the Irish potato blight. Monocultures, which lack the genetic diversity to respond to changing environmental threats, are sitting ducks for parasites. Mr. Currie felt there had to be a parasite in the ant-fungus system. But a century of ant research offered no support for the idea. Textbooks describe how leaf-cutter ants scrupulously weed their gardens of all foreign organisms. "People kept telling me, 'You know the ants keep their gardens free of parasites, don't you?'" Mr. Currie said of his efforts to find a hidden interloper.

But after three years of sifting through attine ant gardens, Mr. Currie discovered they are far from free of infections. In last month's issue of the Proceedings of the National Academy of Sciences, he and two colleagues, Dr. Mueller and David Mairoch, isolated several alien organisms, particularly a family of parasitic molds called *Escovopsis*.

Escovopsis turns out to be a highly virulent pathogen that can devastate a fungus garden in a couple of days. It blooms like a white cloud, with the garden dimly visible underneath. In a day or two the whole garden is enveloped. "Other ants won't go near it and the ants associated with the garden just starve to death," Dr. Rehner said. "They just seem to give up, except for those that have rescued their larvae." The deadly mold then turns greenish-brown as it enters its spore-forming stage.

Evidently the ants usually manage to keep *Escovopsis* and other parasites under control. But with any lapse in control, or if the ants are removed, *Escovopsis* will quickly burst forth.

Although new leaf-cutter gardens start off free of *Escovopsis*, within two years some 60 percent become infected. The discovery of *Escovopsis*'s role brings a new level of understanding to the evolution of the attine ants. "In the last decade, evolutionary biologists have been increasingly aware of the role of parasites as driving forces in evolution," Dr. Schultz said. There is now a possible reason to explain why the lower attine species keep changing the variety of fungus in their mushroom gardens, and occasionally domesticating new ones—to stay one step ahead of the relentless *Escovopsis*.

Interestingly, Mr. Currie found that the leaf-cutters had in general fewer alien molds in their gardens than the lower attines, yet they had more *Escovopsis* infections. It seems that the price they pay for cultivating a pure variety of fungus is a higher risk from *Escovopsis*. But the leaf-cutters may have little alternative: they cultivate a special variety of fungus which, unlike those grown by the lower attines, produces nutritious swollen tips for the ants to eat.

Discovery of a third partner in the ant-fungus symbiosis raises the question of how the attine ants, especially the leaf-cutters, keep this dangerous interloper under control. Amazingly enough, Mr. Currie has again provided the answer.

"People have known for a hundred years that ants have a whitish growth on the cuticle," said Dr. Mueller, referring to the insects' body surface. "People would say this is like a cuticular wax. But Cameron was the first one in a hundred years to put these things under a microscope. He saw it was not inert wax. It is alive."

Mr. Currie discovered a specialized patch on the ants' cuticle that harbors a particular kind of bacterium, one well known to the pharmaceutical industry, because it is the source of half the antibiotics used in medicine. From each of 22 species of attine ant studied, Mr. Cameron and colleagues isolated a species of *Streptomyces* bacterium, they reported in *Nature* in April.

The *Streptomyces* does not have much effect on ordinary laboratory funguses. But it is a potent poisoner of *Escovopsis*, inhibiting its growth and suppressing spore formation. It also stimulates growth of the ants' mushroom fungus. The bacterium is carried by virgin queens when they leave to establish new nests, but is not found on male ants, playboys who take no responsibility in nest-making or gardening.

Because both the leaf-cutters and the lower attines use *Streptomyces,* the bacterium may have been part of their symbiosis for almost as long as the *Escovopsis* mold. If so, some Alexander Fleming of an ant discovered antibiotics millions of years before people did.

Even now, the ants are accomplishing two feats beyond the powers of human technology. The leaf-cutters are growing a monocultural crop year after year without disaster, and they are using an antibiotic apparently so wisely and prudently that, unlike people, they are not provoking antibiotic resistance in the target pathogen.

In a loose team, the four biologists involved in the new findings are seeking to understand the deeper intricacies of the extraordinary system. Dr. Schultz, now at the Smithsonian Institution in Washington, D.C., is a specialist in the evolutionary relationships of ants, and Dr. Rehner, of the University of Puerto Rico, is an expert on fungi. Dr. Mueller, now at the University of Texas at Austin, is an ecologist. Together with Dr. Currie, they hope to unravel the relationships woven between the four members of the symbiosis—the attine ants, their mushroom, *Escovopsis* and the *Streptomyces* bacterium. There are doubtless other members to be discovered.

The four researchers are pleased but not much surprised to have discovered so much new about an already well-studied system. "It may be one of the best studied symbioses in biology but that is a sad reflection on how little we know in general," Dr. Schultz said.

Discovery of the deadly *Escovopsis* fungus could change the correlation of forces between ants and people. In the 10,000 years since the ants have had to deal with human agriculture, a blink of an eye in their 50 million year farming history, the ants have generally prevailed. They thrive in disturbed ground, and so have usually benefited from the clearing of forests.

For the most part, people and the attine empire have co-existed peacefully: the two species live on different scales and in separate spaces, people above ground and attines below it. But attines are a serious agricultural pest in much of the tropics. Mr. Currie, having shown that nests in the lab are devastated with a squirt of *Escovopsis* spores, suggests the fungus might prove a useful way of controlling attine nests.

There seems little objection to applying *Escovopsis* nest by nest. But Dr. Schultz expressed horror at indiscriminately wiping out all fungus-growing ants, many species of which are harmless.

That an *Escovopsis*-like discovery could tip the scales too far toward the human side was foreseen by Dr. Wilson and his colleague Bert Holldobler. In their book *The Ants,* published in 1990, they wrote that biologists need to search for "the weak points" in the ants' social system: "The goal, however, should be intelligent management of their populations and never their complete eradication. Our advantage—and responsibility—lies in the fact that we can think about these matters and they cannot."

—Nicholas Wade, August 1999

In Death-Defying Act, Butterfly Thrives on Poison Vine

THE RAP SHEET could scarcely be more shocking. Poisoning with cyanide. Cannibalism. Molestation of minors. But wait, there are extenuating circumstances. The perpetrator is highly intelligent, with a brain more than twice as large as normal. He is remarkably elegant, if that's an excuse. He leads an unusually industrious life. And he is not a party to this society's moral code, being but a butterfly with perhaps one of his own.

Heliconius butterflies live in the tropical forests of Central and South America aside from one species, *Heliconius zebra*, which has taken a liking to Florida. The adults have distinctive rounded wings with colorful stripes. Most butterflies fly at full speed to escape predators. But heliconians saunter from flower to flower, flaunting their bright hues, as if they hadn't an enemy in the world. The adult butterflies favor flowers known as forest cucumbers but as caterpillars they feed on the leaves of a family of vines that produce one of the most gorgeous flowers next to orchids, the passion flowers.

In the forest, everything that is strikingly beautiful is sinister. The creatures that dare display themselves have hidden defenses. The *Heliconius* butterflies can afford to be nonchalant because their tissues are stuffed with cyanide.

A curious fact is that the passion vines on which they feed are also packed with cyanide-laden chemicals known as cyanogens, although of a different kind than the cyanogens the butterflies make. As soon as hungry mouths start chewing the vine's leaves, its cells release an enzyme that breaks out the cyanide. Cyanide is a universal poison that interferes with cells' respiration.

A common interest in the poisoner's art is just one of many links that bind the insect and the vine together. Dr. Lawrence E. Gilbert of the University of Texas has been studying *Heliconius* and passion flowers since 1969 and has explored many features of their interactive biology. He and two colleagues, Dr. Helene S. Engler and Dr. Kevin C. Spencer, have now begun to work out the chemical interaction between the two. They reported in *Nature* last month that one species of the butterfly, *Heliconius sara*, can convert the principal cyanogen of its food plant to a source of nutrition.

Biologists suspect that the vigorous interaction between the vine and the butterfly over millions of years may have been a driving force in each group's evolution. There are some 600 species of passion vine in the New World, and 40 species of *Heliconius*. It may be that the vines diversified first, buttressed by their cyanide defense mechanism against the usual plant eaters.

Then the butterflies broke the cyanide code and found they had a rich range of food plants more or less to themselves. "The standard pattern is that a plant develops a novel chemical defense and does well for a while, diversifying, and then some lucky group of insects is able to overcome that defense, which allows them to undergo a parallel radiation," said Dr. Andrew V. Z. Brower, an entomologist at Oregon State University.

Even though its prime defense was breached by the butterfly, the passion vines fought back in other ways. Dr. Gilbert has found their leaves are very variable in shape, as if to disguise themselves from hunting heliconians.

Most remarkably, some species of vine have developed a defense system of faux eggs. Female butterflies, if they see what appears to be an egg on a new stem, will avoid it and lay elsewhere. The false eggs, essentially yellow bumps on a leaf, exploit a dark feature of *Heliconius* butterfly: its caterpillars are cannibals. Since they eat only the vine's young shoots, which are large enough to support only one larva to maturity, the first caterpillar to hatch will eat all the other eggs on its shoot.

The vines have developed the faux eggs by adapting another defense mechanism, nectar-dispensing glands that encourage ants to take up residence on the leaves and repel intruders.

The vines may be smart but the heliconians, to hear Dr. Gilbert describing his favorite insect, are veritable Einsteins among lepidoptera. Most butterflies lead what may seem pretty pointless lives: they munch away furiously as caterpillars, hoping to turn into pupae before birds or the dreaded ichneumon wasp discover them, but live just a few days as adults. Apart from sipping nectar, they are incapable of feeding.

The heliconians have transcended this pointless life through a simple but revolutionary advance. The butterflies have learned to feed on pollen by sucking it into their tongues and secreting an enzyme that digests the nutritious grains. This means they can live for months. It also allows them to transfer much feeding effort forward from the vulnerable caterpillar stage to adulthood.

If they can find enough pollen, female butterflies will keep laying eggs until their skin and wings fray. Not for heliconians the nectar-sex-and-hedonism routine of other butterflies. These are busy insects. They stake out home territories, which they patrol every day, checking out every vine in the neighborhood. The females look for any new shoots produced by the passion vines. Both sexes search for the brief-lived flowers of another sort of vine, the forest cucumber, which provides the pollen.

"They are very effective at learning their habitat," Dr. Gilbert said. "Pollen feeding gives them a long life span, so once they find a host plant they can relocate it. You see *Heliconius* butterflies cruise in as if they know exactly where they are going. So we'll see the same individual butterfly checking out the same host plant for months."

Though no one knows too much about the organization of insect brains, Heliconines have a larger than usual "mushroom body," a region of the brain thought to be involved with memory.

The male butterflies have put their minds to at least one reprehensible purpose—cradle-snatching. They have even invented a new crime, which might be called cradle rape. The butterflies are so eager to find mates that they locate female pupae and perch on them, sometimes four males to a pupa, until the female emerges, obliged in her first instants of adult life to manage mating at the same time as she is trying to unfurl her wings.

Some species of heliconians have taken this aggressive suitorship to a further extreme: they probe into the pupa and fertilize the female before she has even emerged. These attentions are lethal for females of heliconian species that do not practice pupal mating, and have the useful result of preventing other species from sharing the same food plant.

Pollen feeding has been exploited by heliconians for more than just nutrition. Long ago, maybe, the butterflies learned how to convert the amino acids from the digested pollen into chemicals known as aliphatic cyanogens. Dr. Spencer, a plant toxin chemist who is now senior scientific adviser for the Safeway supermarket chain in Britain, said this might have been the breakthrough that enabled the butterflies to feed on the otherwise poisonous passion vines. In developing their own cyanogen metabolism, they were somehow able to neutralize the different kind of cyanogens produced by the vines, Dr. Spencer suggested.

If so, this may explain another key step in the butterfly's evolutionary past, its development of mimicry. The heliconians may first have mimicked a poisonous species in a related family of butterflies without being poisonous themselves, an imitation known as Batesian mimicry.

Once they learned how to brew cyanide from pollen, they made their own tissues toxic and developed the higher art of Mullerian mimicry, in which poisonous species resemble one another. Many Heliconine species exist as pairs that mimic one another. The presumed advantage of Mullerian mimicry is that fewer individuals are expended in teaching predators the virtues of avoidance.

While evolving the narcissistic art of mimicking one another, the heliconians have also been engaged in a furious evolutionary arms race with the passion vines. The vines have evolved a witch's brew of different cyanogens to fend off the butterflies, and each new chemical may have required the heliconians to develop a new stratagem to cope with it. One species of passion vine, *Passiflora auriculata,* produces no fewer than five different cyanogens, a perilous cocktail for the caterpillars of *Heliconius sara,* the species that has learned how to eat the plant.

Dr. Engler has now found that the caterpillars can detoxify the principal cyanogen, a complex chemical called epivolkenin, and convert its cyanide into useful nitrogen. Each step in this intricate chemical battle, Dr. Spencer says, has evolved as a tit-for-tat escalation between each species of heliconian and the passion vine species that is its host plant. This relentless chemical warfare may have been a factor in the present profusion of species on each side.

Dr. Gilbert, who has spent 30 years studying the visible biology of the Heliconines and passion vines, hopes a whole new level of interaction will emerge from study of the chemistry and genetics of the system.

—Nicholas Wade, August 2000

Honeybee Shows a Little Gene Activity Goes Miles and Miles

MIDWAY THROUGH A HONEYBEE'S fleeting, bittersweet, and, yes, busy little life, a momentous transformation occurs: the 2-week-old worker must abandon her cloistered career as a hive-keeping nurse, and venture out into the world to forage.

She must learn to navigate over great distances at 12 miles per hour, select the finest flowers, assemble bits of pollen and droplets of nectar into a load nearly as heavy as she is, and then find her way back home. Once there, she must convey the coordinates of her discovery to her sisters in the classic cartographic waggle, the bee dance.

And all this behavioral complexity is packaged in a brain no bigger than the loop of a letter b printed on this page.

Now researchers have identified a crucial genetic component of the great bee leap from homer to roamer. They have discovered that just before the transition, the activity of a gene aptly named the foraging gene increases sharply in the parts of the bee brain that absorb and interpret visual and spatial information.

That molecular surge is clearly the key to the vocational switch. When the scientists fed young bees sugar water laced with a drug that stimulated the foraging gene prematurely, the bees assumed their hunting post far ahead of schedule.

The new research builds upon previous work in fruit flies, and demonstrates that, at least among insects, relatively tiny shifts in gene activity can have striking effects on behavior. The work also suggests that one way to explore the combustible field of behavioral genetics is by looking at creatures that possess an impressive array of skills, as bees do, yet that are sufficiently far from human beings to discourage anybody from glibly misapplying the results.

Dr. Gene E. Robinson of the University of Illinois at Urbana-Champaign and his colleagues reported their findings in the April 26 issue of the journal *Science*.

"There are very few examples of complex behaviors being tied to a single gene," said Dr. Thomas Insel, director of the Center for Behavioral Neuroscience at Emory University in Atlanta. "This happens to be one of the coolest."

Dr. Fred Gould, a professor of entomology at North Carolina State University in Raleigh, said: "When you're looking at physiology, it can be pretty obvious if a genetic change affects renal function, or causes blindness. But behavior has seemed more amorphous." The new work, he said, "gives me a lot of hope that we'll be able to make progress in understanding the genomics of behavior."

How far that progress will extend remains to be seen, of course. Dr. Robinson said that while there was an equivalent of the foraging sequence in the DNA of humans and other mammals, "one thing we are not talking about here is the shopping gene."

Dr. Robinson and his colleagues were inspired to examine the foraging gene in bees by recent results from the laboratory of Dr. Marla B. Sokolowski at the University of Toronto. The Canadian team discovered that there were two alleles, or variants, of the foraging gene in fruit flies, one more active than the other. Flies inheriting the active form developed into so-called rovers, flitting about widely in search of food, while those bestowed with the less vigorous forage allele matured into sitters, content to lounge around and eat whatever fruit was in the immediate vicinity.

Bees and flies are separated evolutionarily by 300 million years, yet Dr. Robinson saw the two fly styles as rough analogies for the two stages of being a bee. As nurses responsible for cleaning up the hive and brood care, bees were like sitter flies; when the bees turned to foraging, they became like rover flies.

Isolating the bee version of the foraging gene (and finding only a single variant of it), the Robinson group determined that the gene affects bee behavior, not from the start, as it does with flies, but in step-wise fashion.

For the first two weeks of bee life, the gene is relatively silent. For the second two weeks of its life, the gene is busy indeed, activated with particular vigor in the optic lobes and the so-called mushroom bodies of the bee brain. The mushroom bodies, named for their resemblance to the fungus, are the main centers for processing a multitude of sensory signals, including those involved in sight, balance and orientation—exactly the faculties that a bee must call on in her new responsibilities as breadwinner.

To demonstrate that the upswing in foraging gene activity was not simply a result of the bee's growing older, but instead was linked to the new behavioral demands, the researchers manipulated hive population dynamics. They removed the foragers, leaving only nurses behind. With no bees bringing home the essential ingredients for honey, some of the nurses turned to foraging precociously. And though they were only days old, rather than weeks, their foraging gene had snapped to life.

Similarly, the researchers were able to recruit nurses to the foraging trade early by feeding them a compound that stimulated the foraging gene.

The researchers have a basic idea of how the foraging gene operates. Using its instructions, a cell generates a protein called a protein kinase, which plunks little molecular bundles onto other proteins in the cell, altering their shape and thus their function. Hence, the real question about the foraging gene is, what proteins does its specified protein interact with? And how does all that molecular commotion end up putting the buzz in a bee's bonnet?

One way to begin sorting out the many genetic players in a bee brain, as well as a bee body, Dr. Robinson said, is through a bee genome project, deciphering all the subunits of bee DNA as has been done, more or less, with the human genome.

The National Institutes of Health is now deciding which genomes of which species are worth decoding in their entirety. Among the candidates are chickens, dogs, sea urchins, cows, freshwater turtles and an assortment of nonhuman primates.

Dr. Robinson believes in his bees. "With a honeybee, you get a lot of bang for your buck," he said. "It has a modest-sized genome, yet it has sophisticated behavior, it learns, it's highly social and because it's an insect, there's a possibility we can understand it in great depth."

—NATALIE ANGIER

Wasp Invades a Spider and Puts It to Work

THERE ARE FEW THINGS more creepy than alien possession, the notion of one creature taking over another's body and bending it to different purposes. Though this may happen every day on other planets, an egregious example has come to light on earth too, and as close to home as the forests of Costa Rica.

Here lives an orb-weaving spider, so called because of the perfect roundness of the web it industriously rebuilds every day. A serious hazard of the spider's busy life is that it is hunted by an ichneumon, or parasitic wasp. If the wasp's attack is successful, it temporarily paralyzes the spider and lays an egg on the tip of its abdomen, where it is out of reach. For two weeks the spider spins its web and catches insects every day as if nothing were amiss, except for the growing larva that clings to its belly and sucks the juices that drip through small punctures it makes in the spider's body wall.

So far this is just the usual grim script of parasitism. But then comes a strange twist. The night before the wasp larva kills its host, it somehow induces the spider to build a most unusual web.

Instead of its delicate orb, the zombified spider constructs two stout silk cables with thick cross-braces in between. This durable platform stands up to wind and rain better than the spider's ephemeral web. The wasp larva then kills the spider, and spins its cocoon on the platform constructed for it, safe from the ants that patrol the ground below.

Dr. William G. Eberhard, the spider expert at the University of Costa Rica who first noticed the bizarre phenomenon, says the wasp larva manipulates a particular subroutine in the spider's web-building program. The subroutine, an early part of constructing the web's frame, usually has five steps. Under the larva's direction, the spider just performs the first two steps over and over again.

This reprogramming of the spider's routine is presumably achieved by some chemical the wasp larva injects into its host. Dr. Eberhard has found that if he removes the larva from the spider on the final evening, the spider will build a platform-style web that night and the following night, but will revert to making its usual orb thereafter, as if recovering from some strong drug.

He has no idea what the chemical might be, but hopes first to identify what gland in the larva may be secreting it. The wasp's behavior is described by Dr. Eberhard in a report in last week's *Nature*.

Dr. Jay Rosenheim, an expert on parasitic wasps at the University of California at Davis, said many parasites were known to shape their host's behavior in various ways. "But what is really amazing and wonderful about this example is that the host's behavior is manipulated in such a detailed way," he said.

As another example he cited the case of a wasp, *Cotesia glomerata*, that parasitizes the large white butterfly. When the larvae emerge from the caterpillar, they spin their cocoons right next to it, and the stricken insect then stands guard by weaving a web over the cocoons and threatening attackers, behavior that is presumably induced by the larvae.

The Costa Rican wasp discovered by Dr. Eberhard preys on an orb-weaver called *Plesiometa argyra*. The wasp, a new species, is awaiting a name from Dr. Ian Gauld, a wasp taxonomist at the Natural History Museum in London.

Dr. Gauld said this was the first time he had seen a wasp with the ability to manipulate its host's behavior in such a way, and that nothing was known about how the manipulation was accomplished.

"I think biology is one of the last great frontiers," he said. "We have got no idea about what there is on earth with us, let alone what it is doing or how it does it."

—NICHOLAS WADE